The Cattle of Great Britain
Various Breeds of Cattle of the United Kingdom, Their History, Management, Etc

by Charles John Darent Blake Coleman

with an introduction by Jackson Chambers

This work contains material that was originally published in 1875.

This publication is within the Public Domain.

This edition is reprinted for educational purposes
and in accordance with all applicable Federal Laws.

Introduction Copyright 2018 by Jackson Chambers

Self Reliance Books

Get more historic titles on animal and stock breeding, gardening and old fashioned skills by visiting us at:

http://selfreliancebooks.blogspot.com/

Introduction

I am pleased to present another title in the "Cattle" series.

The work is in the Public Domain and is re-printed here in accordance with Federal Laws.

As with all reprinted books of this age that are intended to perfectly reproduce the original edition, considerable pains and effort had to be undertaken to correct fading and sometimes outright damage to existing proofs of this title. At times, this task is quite monumental, requiring an almost total "rebuilding" of some pages from digital proofs of multiple copies. Despite this, imperfections still sometimes exist in the final proof and may detract from the visual appearance of the text.

I hope you enjoy reading this book as much as I enjoyed making it available to readers again.

Jackson Chambers

PREFACE.

A SERIES OF ARTICLES descriptive of the various breeds of British Cattle appeared in the pages of *The Field* newspaper during the years 1872 and 1873, and, as they were supplied by eminent authorities on their respective subjects, it has been thought desirable to present them in a collected form, with the original illustrations from the facile pencil of Mr. Harrison Weir, which were, in nearly every instance, sketched from life, the remainder being copied from photographs or reliable drawings. In republishing these articles, however, it has been considered that their utility would be enhanced by the addition of some chapters treating generally of cattle management, seeing that there is a great dearth of accessible information on this important subject, and the "Treatise on Cattle" written forty years ago by the late Mr. Youatt, and justly regarded as a standard work, is now, of necessity, somewhat out of date. Accordingly, the first half of the present volume has been prepared, partially from articles published in the above-named journal, but with copious additions and emendations, in order to bring them up to the present level of agricultural science and practice.

The Editor avails himself of this opportunity of offering his best thanks to all who have so contributed, for the courteous readiness with which they have on all occasions afforded him the requisite information.

J. COLEMAN.

RICCALL HALL, YORK.

CONTENTS.

PART I.—THE GENERAL MANAGEMENT OF CATTLE.

CHAPTER		PAGE
I.	INTRODUCTORY	1
II.	BREEDING AND GENERAL MANAGEMENT	7
III.	PRINCIPLES OF FEEDING—NATURE AND VALUE OF DIFFERENT KINDS OF FOOD	18
IV.	BUILDINGS, AND THE MANUFACTURE OF MANURE	35
V.	DAIRY MANAGEMENT, THE MILK TRADE, &c.	49

PART II.—THE GENERAL BREEDS OF CATTLE.

ENGLISH GROUP.

CHAPTER		PAGE
VI.	SHORTHORNS. By John Thornton	67
VII.	HEREFORDS. By Thomas Duckham	72
VIII.	DEVONS. By Capt. Tanner Davey	78
IX.	THE LONGHORNS. By Gilbert Murray and the Editor	81
X.	SUSSEX CATTLE. By A. Heasman	88
XI.	NORFOLK AND SUFFOLK RED POLLED CATTLE. By Thomas Fulcher	92

PREFACE.

SCOTCH GROUP.

Chapter		Page
XII.	Polled Galloway Cattle. By Gilbert Murray	97
,,	Polled Angus or Aberdeenshire Cattle. By "Scotus"	101
XIII.	The Ayrshire Breed of Cattle. By Gilbert Murray	105
XIV.	West Highland Cattle. By John Robertson	110

WELSH AND IRISH GROUP.

Chapter		Page
XV.	The Glamorgan Breed of Cattle. By Morgan Evans	117
XVI.	Pembrokeshire or Castlemartin Cattle. By Morgan Evans	122
XVII.	The Anglesea Cattle. By Morgan Evans	129
XVIII.	The Kerry Breed of Cattle. By R. O. Pringle	134

CHANNEL ISLANDS GROUP.

Chapter		Page
XIX.	The Alderney Breed of Cattle. By "An Amateur Breeder"	139
XX.	The Breton Breed of Cattle. By J. C. W. Douglas and Others	146
XXI.	The Guernsey Breed of Cattle. By "A Native"	152

THE CATTLE OF GREAT BRITAIN.

PART I.
THE GENERAL MANAGEMENT OF CATTLE.

CHAPTER I.

INTRODUCTORY.

HORNED CATTLE have been cultivated in this country from the earliest times. We are told that the Britons neglected the art of cultivation, so well known to the Romans, and contented themselves with looking after cattle, living on their flesh and milk. No doubt breeds were kept up by a process of natural selection. Little care would be bestowed upon the selection of sires, but the stronger animals would be reserved as males, and, running out with the cows, lived in a condition of semi-wildness, of which we see instances now in the white cattle of Chillingham Park. It is probable that the descendants of the original cattle are those which we see in Sussex, Devonshire, Wales, and Scotland; and it would not be difficult to trace a certain likeness. They are all middle horned type, but climate and food have doubtless caused great changes. The Longhorns, which were originally derived from Ireland, first took root in Lancashire, and from thence spread to the midland counties, where for a time they formed the prevailing breed, being superseded by the Hereford, which probably were derived from the same stock as the Devon. The Shorthorns are evidently of very mixed origin, possibly owing some of their merits to foreign blood. They were at first restricted to Yorkshire and Durham, but have, from their superior qualities as rent-paying cattle, pushed their way in all directions, and become established wherever climate and soil are sufficiently good.

Different breeds differ considerably in aptitude to feed, the more cultivated sorts —those which differ most from the original types—are as a rule the quickest feeders.

Thick-skinned animals are proverbially slow, although the quality of flesh is good because more compact. The less highly cultivated sorts thrive upon harder keep, and can be kept with a profit when the better sorts would starve. And this to a certain extent limits the utility, and points out the localities best suited for the different breeds. The Shorthorn above all other breeds exercises an influence on quality. Ireland affords the most telling example of this fact. Only a few years ago Irish was a term of reproach as applied to cattle, and the hard-skinned big-boned mongrels exhibited in our fairs were distinguished at a glance from home-breds. Now, thanks to the prepotency of Shorthorn sires, the best lots of Irish are quite equal in appearance, and often higher in price, than good stock bred here. The marvellous change that has been made proves how great the value of good sires, and how important it is to use well-bred and well-shaped animals. But it is not necessary to go to Ireland for examples of Shorthorn influence. Compare the character of the stock in any of our markets now with what it was thirty years since. The change is entirely attributable to the influence of good bulls.

Whoever travels through the length and breadth of this country cannot but be struck with the general quality of the cattle which everywhere meet his eye, and if he can look back for thirty or forty years the progress that has been made will appear in a very favourable light. North, south, east, and west, with few exceptions, it is the same, some counties being better off in this respect than others, those in which the holdings are the largest being the best. But, although much has been done in the way of improving our breeds of cattle, there still remains a very great deal to do. For example, in several of our principal dairy countries the quality of the cows is not as good as it might be; and as it is from these districts that much of our store stock comes, it is through their medium that we must look for an improvement in the general quality of the stock spread over the surface of the country. Too often the cows kept have no tendency whatever to make flesh, even when dry and well fed. No amount of good food could render such animals fat, and their offspring must inherit their bad properties. An idea prevails that by improving the quality and meat-producing properties of dairy cows their value as milkers will be lessened; but this need not be the case if proper care be taken in selecting bulls of well-known milking families to cross with the existing stock. There are, even among the very highest-bred Shorthorns, cows which give as much milk and butter as common ones; and where milking powers have not been disturbed by unnatural feeding, and due care has been exercised in the choice of bulls, Shorthorns yield more milk than any other pure breed. It should also be kept in mind that we breed oxen as well as cows.

If a farmer buys a bull simply because it has a pedigree and is a bull, without any reference to the milking properties of the family from which he springs, such a man does not deserve to be lucky in his purchase. Farmers should take more pains to inquire into these matters before purchasing, and then breeders of bulls would find it conduce to their benefit to pay more attention than they do now to the milking properties of their stock. The results of the use of high-bred bulls with rough cows are sometimes astonishing, the calves partaking so much of their sire's quality, being smarter looking, and having far better coats than their mothers.

In these days, when it has become so much the fashion to rear calves—and more especially has this been the case since the cattle plague decimated our herds—well-bred, coloury calves meet with a ready sale at very remunerative prices. In some places a good cow calf a week old will fetch 3*l*., and bull calves 2*l*. Everyone who has had any experience in rearing must know that a half-bred calf is much dearer at 1*l*. than a good one at 2*l*. When they both are a year old, the one will be a very different animal from the other.

We must impress upon farmers generally, and young ones in particular, the importance to be attached to colour in choosing animals to breed from. More stress should be laid on this point than would at first appear, so-called coloury stock finding a much readier sale than when the predominating colour is white. It may be prejudice, but graziers do not like white stock, and will not have them if possible. They are not considered so hardy as reds and roans, and as store stock, especially if it is intended to do them roughly, they have many disadvantages. It is a fact that white animals seem more liable to parasites than their richer-coloured companions. How this is to be accounted for we cannot vouch, but there can be no doubt as to its truth. Cattle will be found to deteriorate, so to speak, in colour—*i.e*, they will, generation after generation, become lighter. In order to obviate this, it is necessary to keep the colour up to the mark by the introduction of fresh blood of the desired colour, and *descended from stock of a similar colour*. Animals exercise influence on their offspring in proportion to the purity and *length* of their descent; therefore, a pure-bred white bull will be likely to beget more white calves than coloured ones, although all the cows put to him were red ones, always supposing the cows not to be of equal purity of blood. If we start with red as a foundation colour we can easily get the stock lighter, and therefore this colour will be found the best. The use of red bulls bred from red stock will, in a short time, influence very greatly the colour of a herd.

It is a mistake to suppose that well-bred animals require more food, and that of better quality, than rough ones. No doubt a Welsh runt will do well where an average-bred beast would comparatively starve; but between the common black-nose hard-skinned, light-fleshed cow so often seen, and one with two or three crosses of pure blood, there is no difference whatever in this respect. We will not say that the better-bred cow will do *better* on *very poor* food, because we believe such not to be the case; but if the quality of the food be improved, then the well-bred animal will soon show a marked superiority.

Store stock are now very dear, and as the supply will always be limited, there is every prospect of a continuance of good prices for this kind of stock. As long as this is the case the difference in the value between good and bad stock, when they come to be sold out, either barren or in calf, cannot but be of vital importance to the farmer. In the case of poor-bred, barren cows there might be very considerable difficulty in selling them at all, whereas in the other they will command a very good price indeed, a really good cow being often worth almost as much barren as in calf.

We have endeavoured to explain how it is possible easily and speedily to enhance the value of a herd without being at any great expense. Shorthorn bulls may be bought at a moderately low price as yearlings, or at a more mature age, either at sales or by private contract. At sales the purchaser, unless he have a previous knowledge of the herd, must remain more or less in the dark in regard to their qualities and their probable utility for his purpose. Against this he must set the chance of securing a cheap bargain. Bulls of from four to six years old may often be bought at a butcher's price, and, providing they will get stock, will often answer the dairyman's purpose better than a younger one. For this reason, young animals of fashionable blood will always command good prices, whereas older ones will be sold cheap. It must not be inferred, because a bull has failed to get show animals from high-bred cows, that he will not answer when used upon rough ones. The better his quality the more marked his effects.

The Shorthorn, with its capacity for early development, combining, when properly selected and carefully bred, milking and feeding properties, yielding for a given quantity of food a larger return than any other breed, is the animal that seems to offer the greatest advantage to the breeder, and it is not exceeding the truth to say that when circumstances are suitable the Shorthorn will be patronised. The very fact that, originating some ninety years since on the banks of the Tees, in a comparatively small district, they have become distributed through the length and breadth of the land, and penetrated to the far distant shores of America and Australia, speaks volumes for the merits of this breed. Ireland, once remarkable for its mongrel stock, owes the vast improvement of its herds to Shorthorn influence. In parts of Scotland the Shorthorn is to be found, and if he does not displace the aborigines, he blends with them and produces magnificent cross-breeds. The Shorthorn, however, requires good and abundant food, and is not suited to exposure in a severe or very moist climate. The young animals must be well cared for, as after-growth greatly depends upon a supply of nutritious food during early life. The influence of good bulls in our dairy districts has been very marked; we have heard it said by a successful breeder in Gloucestershire that twenty-five years since, when he commenced breeding, there was not a Shorthorn within many miles; now it would be an exception to find a herd that is not three parts pure, and many of the farmers possess pedigree stock. The animals that have been displaced were great at filling the cheese-tub, but so coarse and slow-feeding that, although the produce may have been to some extent sacrificed, the total return is better and quicker. The draft cows, which formerly were sold poor at low rates, are now finished off with a little cake and roots, and command good prices. The young stock also sell well—calving-down heifers making 25l. to 28l. at three years, and steers coming out fat at the same figure, when between two and three years old. Shorthorns are more subject to sterility than less cultivated breeds. This arises from the unnatural condition in which high-bred animals are too frequently reared, and may be guarded against by giving our young stock plenty of exercise and keeping them from fat-producing food. The researches of physiologists have demonstrated that excessive fatness of the carcase is accompanied with deposition of fat in the tissues, and when this is the case the breeding tendencies in both sexes are seriously compromised. Wherever the stock is treated naturally, and the bulls selected with care, barrenness is not present in the Shorthorn to a greater extent than in other cultivated breeds.

The Herefords, a much older breed, have never become so thoroughly scattered although they have pushed their way towards the circumference of a wide circle, and have displaced or become mixed with the native breeds. Thus we find them in Shropshire, Warwick, Stafford, Monmouth, and several of the Welsh counties. Great attention has been paid to this breed of late years, and in many points of general utility the Herefords are unrivalled. The quality of the beef, and the capacity of the beast to lay it on rapidly and in beautiful proportion of fat and lean, are remarkable. The springy firm touch of a well-fed Hereford is due to the distribution of the fatty globules. The Hereford is very hardy, capable of doing upon poorer fare than the Shorthorn, not requiring such careful attention whilst young, and thriving in exposed situations. The pasture lands of the old red sandstone suit them well; here they rear their produce, but are not, as a rule, large milkers; perhaps the tendency to make beef is too prominent to allow of large dairy produce. A well-bred heifer or young cow will, on good food, fatten whilst milking, and no feeding stock will pay so well as Hereford cows about five years old. The great object in the district is to rear the produce and bring it forward, so as to sell either as yearlings or two-year-olds; they are bought for grazing principally by the farmers of the midland and eastern districts. Formerly a good trade existed for working bullocks; but latterly the great value of young beef has caused everything good to go that way, and it is only the inferior lots, and very few of them, that go into harness. The Hereford is a hardy healthy animal, with many valuable qualities.

Wherever the climate or food approaches to the character of our mountainous districts, we shall find that local breeds are the most desirable to keep. We may improve them by care, or even by judicious crossing; but we cannot dispossess them without certain loss. Thus the North Devon cattle are admirably adapted, from their active habits and hard nature, to feed over the exposed ranges of Exmoor and similar districts; whilst both in Wales and Scotland we find distinct breeds modelled, as it were, by the force of circumstances, into forms best suited to withstand the climate. Still in many of these cases, as our system of management improves, we may be in a position to make use of a more cultivated animal, and either improve the originals by careful selection and good treatment, or try the effect of crossing.

We have said enough to show that the subject requires general rather than particular treatment. At the same time, as it is necessary to have some type of animal before our eyes, we shall describe the management suitable to any of the more cultivated breeds, and especially to animals of the Shorthorn type.

We do not propose to treat of high-bred animals, as such are not generally desirable for the rent-paying farmer; or, at any rate it is not wise to commence by a heavy outlay in cows with long pedigrees, unless we have time, taste, pluck, and money to go in for breeding prize stock—a more interesting than profitable affair with most. Good-looking roomy animals, got by a pure-bred bull out of ordinary cows, or animals whose length of pedigree does not materially affect their price, must be sought for. It is bootless to describe those points that indicate the dairy animal—such knowledge can only be acquired by experience and observation; and the young farmer may very reasonably doubt his own judgment, and will do well to commission a respectable dealer, who is generally able to make a more profitable selection and obtain the animals on better terms than the farmer himself: the practice of employing a middle man, both for buying in and selling out, is increasing in our grazing districts.

CHAPTER II.

BREEDING AND GENERAL MANAGEMENT.

A GOOD lot of young cows having been secured, the next point is to select a bull, and the wiser policy is to obtain a thoroughly good animal with a sound pedigree, even if we pay handsomely for the same. The first male will have a most important influence on the herd, and a few pounds therefore should not be grudged; generally a good yearling can be had for between thirty and forty guineas.

We are not venturing into an exhaustive treatise on the principles of breeding, or intend more than briefly to touch upon this point, but we must urge upon all young farmers the value of quality, and the improvement of their stock by the use of good bulls. This, be it remarked, is quite apart from keeping highly-bred stock. Pedigree breeding is a business for the few, requiring special conditions to render success even probable; but everyone who breeds—whatever the class of animal he selects—should aim at quality, by which we understand the qualification to mature at the earliest possible period, and accumulate the maximum weight from a given quantity of food. What the difference is in this respect, according to quality, has never been accurately tested, but we believe it is quite sufficient to determine profit and loss. Now, as a rule, the influence of the male preponderates, consequently, whilst careful to select good-looking females, we must spare neither money nor time in finding the right sort of bull. Suppose we require milking stock (and whatever the particular direction in which we farm, milk must always be an important consideration), not only should we select heifers that give promise, but we must seek a sire that comes of a good milking stock, for these qualities are to a great extent hereditary. We must, moreover, take care that the qualities that existed in the ancestry have not been weakened or destroyed by injudicious breeding or feeding. Many an animal with a natural tendency to milk has been ruined by early forcing. Whilst, as we hope to show in the following pages, generous diet from birth is necessary to quick and healthy development; undue forcing, such as is resorted to in order to develope abnormal growth in show animals, weakens and often completely destroys milk-producing qualities. It is in this way, principally, that discredit has been thrown on certain families of Shorthorns, as milk-producers; and thus the race that were originally noticeable for the quantity of their yield, are now frequently unable to rear their produce, and choice animals require foster-mothers to supply their wants.

The folly of forcing young animals for show purposes is acknowledged on all sides, and those who possess the most valuable blood will not run the risk of damaging their animals by forcing. It may be allowable to make an animal extraordinarily fat for the butcher, in order to show the public of what a breed is capable: the animal is for the shambles, and provided he lives till ready for the knife, the end sought for is obtained, and the feeder is the only loser—the extra fat costing more to put on than it will yield; but this extra fat state is not a healthy condition, and animals so fed lose much of their vital energy, as was too evident from the collapse of the beasts at the Smithfield Show of 1873. Lean stock would not have suffered to the same extent. Shorthorns, especially, suffer from the disease known as fatty degeneration of the heart, which may be explained to the unlearned as meaning a deposit of fat between the muscles of the heart, which greatly lessens and sometimes altogether arrests the expansive and contractive powers. The object of the breeder should be to so treat his young animals as to develope frame and flesh, by supplying food containing the constituents of bone and muscle, and allowing of sufficient exercise to develope and strengthen the frame and constitution.

Very little is really known on the subject of breeding. Mr. J. K. Fowler, who delivered a lecture on the subject before the Central Farmer's Club some years since, considers that the sire influences form, the dam the internal organisation, and there is some general truth in this, as is proved by the cases of the Mule and the Hinney. The former, which results from connection of the male ass with the mare, has all the external points of the ass, only a rounder barrel to give room for the bowels, which resemble those of the dam. When a stallion is put to the female ass, the result is a modified horse (the hinney), only the barrel is smaller, in form resembling that of the ass. If this is correct, we should be most careful that the dam is sound in wind and of good constitution, being most particular that the sire possesses beauty of form. In the case of Shorthorns, certain sires, such as Hubback, Favourite, and the Earl of Dublin, were noticeably prepotent in their influence on progeny. The last-named bull, in the hands of Mr. Adkin and Sir C. Knightley, impressed deep milking properties in all the animals he was put upon, and this, a result of internal organisation, is due to the influence of Princess, from which he was directly and closely descended. This latter fact rather tells against Mr. Fowler's theory, since the milking properties derived from Princess were transmitted more directly through the male than the female. Prepotency may arise from an intensifying of certain qualities from very close breedings, and this, also, to some extent, militates against Mr. Fowler's theory. The influence of the male may be due to his being deeper bred than the female. Cross-bred cattle offer a good example. It is nearly always possible to find out the sire by the strong likeness to him, and here we have an illustration in support of Mr. Fowler as to external form. Mr. Fowler further illustrated his views by reference to facts noticed in poultry breeding. The Brahma and Dorking fowls were crossed with the following results: when the Brahma cock was used on the Dorking hen, the chickens had four claws generally, feathered legs; the pullets laid white eggs, and the cockerels, though resembling the Brahma, crowed like the Dorkings. When the process was reversed, the produce likewise followed the change closely; the Cockerels were like Dorkings, but roared like the Brahmas; indeed the illustration was perfect. The same results were noted when the Rouen and Aylesbury ducks were crossed.

Great importance, undoubtedly, is attached to the first impregnation, and the imagination has a good deal to do with colour; and it is said that Mr. McCombie, the celebrated breeder of black cattle, is most careful to have all his buildings, gates, &c., painted black.

The period of the year at which our cows should calve will depend upon circumstances; if our object is rearing and dairying, the calves should drop from Christmas to March or April; if we are for cheese, March and April; and if we are milk-sellers, the cows must come in at all periods. Early calving is best for the offspring. The difference of wintering between a calf dropped in January and May, both receiving equal care, is very great.

As to the best age to commence breeding, different opinions exist, and everything depends upon the class of animals we possess, and the quality of our land to favour early maturity. As the question is a very important one, it would be well if experiments were carried out to determine at what age heifers will breed. If conditions are favourable, the calf that drops in December, January, or February, may be brought to the bull the summer or autumn of the succeeding year, when from sixteen to twenty months old. The first calf will then be dropped when the heifer is under two and a half years. Of course, we presuppose careful attention and abundance of food. The heifer may be small at this time, but grows rapidly afterwards, and we bring our animal into a productive state at the earliest period. The milk may not be very abundant, but there is plenty for the calf, which should be allowed to suck at any rate for some time, as the bag is thereby developed and rendered soft; at six months it is good policy to dry the heifer, as she is thus enabled to lay on flesh and take care of the fœtus. In breeds that are not so forward, or where circumstances are unfavourable for early development, the heifer calves at three years old. In reference to this point, we could multiply examples of accidentally early breeding which have turned out well. So fully satisfied are we, from our own experience, that with generous feeding heifers may calve down when two years old, without injury either to growth or milking qualities, that we adopt the plan of taking a calf from animals intended for beef. Those who buy in much of their stock might carry out this practice. We can often buy, during summer or autumn, yearling Irish heifers; if these were served at once, and done well to through the winter, the plan would answer. Where breeding is carried on, we are quite certain of the profitable nature of this practice. We ascertain at an early period whether the animals are likely to make valuable milkers, and such as are not promising can be fed off after the calf is weaned. An impression prevails that early breeding affects after growth; but we have not found it so where care was exercised as to food. Mr. Edward Bowley, who is well known as an authority on Shorthorn management, alludes to the subject in his prize Essay in the "Journal of the Royal Agricultural Society." His practice is to bull the heifers dropped from December to the end of February, in July or August of the following year—that is when they range from sixteen to eighteen months old—they thus would calve just before going to grass—when they are about two years and four months old. He says: " I allow their calves to run with them during the summer. When four or five months old I take the calves away, and dry the dams, by which means the heifers get a much longer rest than the older cows before they calve again, thereby encouraging their growth; and under this system they can produce calves at an early age without interfering with the full development of their forms." He also mentions a case of very early breeding by a heifer that calved at fifteen months, having been served by a six-months old bull calf whilst both were with their dams. The heifer took the first prize as a two-year-old in-calf heifer, and a second prize the following year as a cow in milk, in a strong class, and was afterwards sold at a high price to go abroad. This is important evidence bearing on the point we are anxious to see elucidated. Our own experience is that, so far from early breeding injuriously affecting future size, the heifer, if generously fed, appears to grow out in consequence, and we are quite certain that for feeding purposes they are the better for having dropped a calf. We have not, however, had so much experience of early breeding for the dairy. Before calving, the cow should be placed in a loose box or shed by herself, and not tied up. It is a most unnatural proceeding to allow cows to calve when tied up in their usual places, among, perhaps, twenty others. The calf runs considerable risk of being injured by the other cows in case it is born when no one is in attendance. The plan is simply cruel, and on no account to be followed. Of course, sometimes such a case occurs on the best regulated farms, but in these instances the cow either calves before her time or unexpectedly. After calving, the cow should have warm gruel and a little sweet hay. Chilled water should be used for the first three days, after which if no unfavourable symptoms occur, all danger ceases. If there is any fear of fever (as is always the case with large milkers), a moderate aperient may be given. Linseed oil is useful for this purpose, and safer than salts and sulphur, which, however, are frequently given. Cleansing drinks should be always at hand, in case the animals do not clean properly.

Great loss is often occasioned by cows slipping their calves. There seems to be no accounting for this; as a rule, the food the animals have been fed on gets the blame, but if anyone will be at the trouble to inquire into the matter, they will find that cows slip calf on all kinds of food, and under all sorts of management. When one cow among a lot of others slips her calf, she should be at once separated from them, and not allowed to be with them again for some time. Unless this is done, and at once, the farmer cannot be sure when it may stop—perhaps not before half his cows have followed the example.

The management of the calf is the next point for consideration. In the case of the heifer it is well to let the calf suck, and possibly run with its dam, during the summer. In such cases we advise the farmer to procure a second calf, and when the heifer has become accustomed to it, to turn her out into good pasture with her two attendants, who will make good use of their time, and pay well for the heifer's milk. With older cows there are two plans open to us—first, to remove the calf at birth, before the cow has noticed or licked it; this plan is frequently pursued in the north of England with great success. The calf is carried to a warm well-littered house, and thoroughly rubbed with a wisp of straw until dry and warm. The beastlings are then drawn from the cow, supplied to the calf a small quantity at a time, and frequently. The fingers should be introduced and the calf's mouth drawn down to the milk. They will thus readily learn to drink, and the great point is to prevent their drinking too fast. We must imitate as much as possible the process of sucking, by which a good deal of air enters the stomach and assists digestion. Patent feeding-mouths are very useful for this purpose. The calf should be fed three times a day; many people prefer only twice, but it is too long for the stomach to remain without food, and is contrary to the natural habits of the calf. The second plan is to leave the calf with the mother two or three weeks, or at least allow it to suck night and morning; but if we have a good cowman who understands the other plan, it is preferable for some reasons. The cow gives her milk down more freely, does not fret at the separation, and is apt to take the bull sooner than when the calf sucks. Adopting the first plan, we may use new milk for a fortnight; then skimmed milk of the same temperature as the new milk, and thickened with linseed jelly or fine dust oil-cake, which supplies the fatty matters removed in cream, besides enriching the food in other ways. Boiled flour porridge is frequently used in conjunction with skimmed milk. We prefer dust cake, provided it be from fresh genuine linseed; great care should be exercised to secure a good article. Care should be taken in all cases that the milk be given warm: cold milk produces scouring, and all manner of evils. The cake may be soaked in hot water first, or else added to the milk and gently heated; in either case it produces a rich soup, which is very palatable and nutritious; a handful of finely-ground oatmeal may be added, and a little later a small quantity of fine pollards. Rock-salt and chalk should always be placed within reach. One great advantage in the plan of separation consists in the earlier date at which the calf eats. As soon as this is accomplished, we may by degrees discontinue the liquid; at first supplying it only once a day, and soon leaving it off altogether. And thus a good cow will rear eight to ten calves, provided her produce is entirely used for this purpose.

When a month old, calves will begin to nibble a little sweet hay, finely-sifted chaff, pulped roots, and meal. We cannot begin too soon to teach them to eat, although the longer they get the skim milk and porridge, the better for future growth. Calves weaned too early seldom thrive well afterwards. Oil-cake, crushed, and then boiled to a jelly, and mixed either with the porridge or skim milk, is excellent food for calves when a month or six weeks old. What we here call porridge, should be called more properly gruel, and should not be made too thick. When ten weeks old, the calf should be weaned; this should be done by degrees, the daily allowance being decreased, so as to accustom it to the change. By this age the calf can eat a considerable quantity of hay, chaff, pulp, and corn, and should receive at least half-a-pound of cake and corn.

The treatment of the calf during its first year is most important. As the spring comes on and the sun gets power, the calves, in small lots, should be allowed the range of a comfortable yard and shed, taking particular care that they are warmly housed at night. If we can depend upon our arable land for a succession of green food, then the calves will do best if kept in yards all the summer, and, indeed, not suffered to go out into the pastures until turned one year old. This is often impracticable, but it is much to be recommended in all cases where it is possible. The advantages of this plan are, that we have the animals more under control, the least malady is at once detected, the supply of food can be regulated with greater exactness, and the manure economised. Considerable variety of food is necessary. Thus, we must have vetches, trifolium, cabbages, and artificial grasses, coming on in regular succession. Chaff, meal, and cake correct the laxative tendency of our green crops. All who are acquainted with rearing know how frequently calves will scour when in the pastures, and how difficult it is to cure this. In many instances we believe the disorder is occasioned by poisonous acrid weeds growing in the grass. Again, as the hot weather comes on, the animals are driven wild by the flies, and while thus irritated cannot make any progress. The yard system relieves us of these difficulties. The green food may be supplied partly long, in racks, and partly cut up and mixed with hay-chaff, a slight fermentation being often desirable. Over this a mixture of several kinds of food may be dusted, amongst which old beans should have a prominent place, being particularly adapted to calves eating green food, counteracting the too laxative effects of the latter, and supplying a large amount of flesh-forming material; we have seen excellent results from the use of beans. A moderate quantity of fine-ground oil-cake, and a little home-grown grain will complete the mixture, unless we can buy bran cheap, in which case it may be added. The quantity of such a mixture to be used depends upon the age and condition of the calf—from 1lb. to 2lb. a day expresses about the range of quantity. With this mixture, continually varied as to the green food, the calves should be fed three times a day at early morning, noon, and night, and clean water always placed within their reach. Occasionally, every two or three weeks, some flowers of sulphur, about half to three-quarters of an ounce per head, may be given with the food; this purifies the blood, and helps to preserve a healthy condition of skin. Judgment and attention are required in supplying the food. Some stockmen fancy that too much cannot be given, and so they fill up the fresh food on what remains from the last meal—a dirty, slovenly practice, that cannot be too severely condemned. An hour or two before feeding time the manger should be thoroughly cleaned out, every particle of dirt removed, and, besides this, every now and then thoroughly washed and scrubbed. Eating off a dirty plate would not tend to improve our appetite, and though the ruminant is not so particular, cleanliness is very desirable. The refuse from the calves' mangers will be readily eaten by the cows, and thus nothing need be lost, and the calves will come to their food with an appetite. Should an animal hang back, we must at once separate it, and carefully watch the symptoms.

There is one point upon which we would specially remark—viz., the importance of keeping the skin healthy; it has certain functions to perform, its excretions relieve the system of waste matters, and there is great sympathy between the skin and the digestive organs; indeed, the lining membrane of the stomach, &c., is but a continuation of the skin. Now, if the pores become choked up with dirt, or if parasites, such as lice, ticks, &c., make their habitation therein, the action of the skin is impaired; the constant irritation frets the animal, and progress is checked, even if more serious injury is not caused. Occasional washing with soft soap and warm water will be found very beneficial, and if parasites are present we may add a portion of carbolic acid, or we may use one of the preparations of this powerful agent which contains a certain quantity of soap, and makes a milky emulsion with water. We are acquainted with herds where this is used, not only for the calves, but through the dairy, with satisfactory results. Moreover, the well-known antiseptic properties of the carbolic acid group render them doubly useful as a possible means of keeping off contagious diseases; and although the evidence is but negative, and must be taken only for what it is worth, we believe that during the cattle plague no disease occurred where this practice of occasionally washing over the animal's body was persevered with. Whether this be so or not, there can be no question about the importance of keeping the skin of the young animal clean, as we thus insure a good circulation, waste matters are duly carried off through the pores, and the internal organs are not overworked. In an ordinary way, two washings in the first year will be sufficient—say, when five or six months, and again at nine or ten months old. During the winter, and especially if there is barley-straw about, we must watch for lice, and wash at once if we find them. It is want of attention to little matters of this kind that so frequently prevents success; our animals contract some irritating skin disease, or are plagued with vermin, and are restless and worried, however well they may be fed.

The plan we have advocated as to summer feeding the calves may not always be possible. In that case we must turn out where there is a good bite of fine herbage, avoiding rank strong pastures. By the end or middle of May, the oldest calves on a dairy farm will probably be three months old and may be turned out by day in a well sheltered grass field, with a shed attached—if they have no shelter to run into they will suffer much both from sun and rain. The most convenient shed for calves is one in which they can be shut up at night, thus saving their having to be driven every night to the homestead. When such a shed is available, by the middle of June the calves may be left to go in or out as they like at night. Care must be taken to have some ground to change them on occasionally, and plenty of good water and a shelter hovel are indispensables. Chalk and rock-salt must be placed within reach, and a supply given night and morning of artificial food and chaff—the latter composed of a mixture of green hay and whatever forage crop we are cutting at the time. Best linseed cake, at the rate of 1lb. per head daily, will answer admirably provided the grass is not very watery, in which case we should prefer using either a portion of beans or decorticated cake, say $\tfrac{1}{2}$lb. of linseed and

½ pint of beans, or ¼lb. of decorticated cotton cake. Common cotton cake should never be used for this purpose, as it cannot be properly digested. Later on in the autumn, whatever the farmer intends to give them during the coming winter in the way of corn, should be gradually introduced and mixed with the oil-cake, &c., so as to accustom them to the change, violent transitions in the way of food are always productive of loss of condition, and therefore, especially with young animals, the change should be gradual.

By such treatment carefully persevered in we shall gain a year in point of bulk and maturity over the old starving system, and have a good chance of ultimate size, which is out of the question when growth has been checked. The advantage of this system of management will be in proportion to the quality of our stock; badly-bred coarse animals do not repay us for our care in the same way as the better sort. Whilst, however, advocating liberal treatment, we must guard against giving an excessive quantity of nutritious food. The young animals should be placed under those conditions that favour rapid growth rather than the accumulation of fat or a great quantity of soft flesh. We like to feel a loose soft hide, with plenty of room under; we must have, in short, a well-covered frame, with a skin that fits on the animal like a loose great coat.

It is not good policy to let calves lie out even during the height of summer, but it is decidedly injurious late in autumn, when heavy dews and frost indicate the chilly nature of the surface. By the 1st of October the calves should be housed for the winter; if allowed to remain out later than this they will not thrive well, the nights becoming cold, and they will also be liable to disease. Of course they may be turned out during the day, but even this we do not recommend for long, as the grass they eat gives them a dislike to the drier food they ought to consume during the hours they are in the yards. They should be gradually brought from grass to dry food and winter management, so as to become accustomed to the change without loss of condition. Sheltered yards, well littered, with good hovels, are the best situations for calves in winter. The smaller the lot in each yard the better. Pulped roots and chaff put together for a short time, to allow of a gentle fermentation, are the materials on which we must chiefly rely. The chaff may be a mixture in equal quantities of oat-straw and hay; over this a small quantity of artificial food may be dusted, in which the greater the mixture of different materials the better—say, barley, wheat, and bean meal, with finely ground cotton-cake or linseed-cake, and perhaps a little palm-nut meal. Yearlings will eat from 1½lb. to 2lb. per day. The cost of such food, taking 2lb. per diem as the average consumption, will not exceed 1s. 6d. a week. The gain in the greater growth and better health of the animal will be very manifest. The natural food goes further, and the manure is improved.

We should not recommend, for calves, the use of any condimental foods which are so loudly advocated by the different makers. As a rule, these foods consist of a mixture of finely ground grain, with a per-centage of locust beans, flavoured and seasoned with certain stomachics and stimulants; the latter being commonly fenugreek, aniseed, carraways, gentian, mustard, and possibly nitre. We could not say a word in favour of such compositions as food for calves, or as the seasoning for their food, because we consider them very objectionable. If the young animal is in a healthy state and under proper management, digestion and appetite will be quite equal to the wants of the body. The use of such food will tend to impair the digestive faculties. We should not consider it desirable to give a child mustard or hot pickles, inasmuch as such stimulating food would injure the stomach; and so the use of cattle condiments must be condemned for young stock, however advantageous in the case of fattening animals. Even if there were not these drawbacks, the condiment, as tending to create an abnormal appetite, would induce the deposition of fatty matter, and thus predispose to disease. To quote from a writer on this subject: " It is now generally admitted by physiologists that an extra development of fat is opposed to the normal growth and health of muscle, and when the accumulation of fat is in connection with any of the vital organs, as the heart, the degree of health is then very low, the slightest influence from without—extra heat or cold, or sudden exercise or excitement of any kind—being sufficient to endanger life. An extra deposition of fat about the liver or kidneys is also attended with similar conditions unfavourable to the general health. High-bred stock too often inherit a tendency to accumulate fat, which all our care is unavailing to counteract. The best antidotes are plenty of exercise, which acts by exciting the excretory organs to free action, with a supply of food suitable for building up the frame; and if, in spite of such treatment, animals will still get fat, we don't know that there is any remedy. Nothing is more opposite to nature than the system of forcing upon stimulating food and keeping young stock shut up in houses—plans that are too frequently adopted in the case of animals intended for exhibition.

Having wintered well during their first year, after-management is comparatively easy. The animal will be able to take care of itself. Our practice must now depend upon the end in view; if we intend fattening the animals, the process must be continuous, and a supply of cake and other artificials must be given at grass, or supplied with the forage crops in the yards; whereas the animals destined for stock may be left to get their own living during summer in the pastures, and, provided they have plenty of food and occasional change, will keep in growing condition, and be quite ready for the bull in the autumn.

Contrast the system we have sketched out, which we know to be based on scientific principles and sound practice, with the absence of management that results in the miserably underfed and underbred animals that may be found at any of our large fairs. Granted that the cost of keep has been only half, are they half as valuable? A good yearling will often make 9*l*. 10*s*. to 10*l*., which is more than a half-starved two-year-old is worth, for the latter are often dear at any price—they have no go in them, no quality that can be developed by keep. It is well sometimes to buy a young animal in poor condition, because we know it will improve rapidly, provided there is quality; but a half-starved mongrel remains a sorry brute to the end of the chapter. Therefore keep a good sort, rear well, and bring to market, or into productive condition, at as early an age as is consistent with the health and stamina of the animal.

Mr. Henry Ruck, of Cricklade, Wilts, described his practice in the management of calves before the Cirencester Farmers' Club, and showed that he had for some years been in the habit of rearing on an average fifty to sixty calves with the produce of four cows. The description embraced three years, and during the first two not an animal was lost. There were three deaths in the third year, but this loss was due to mismanagement. Independent testimony was forthcoming as to the healthy appearance of the stock. The following is Mr. Ruck's description of operations: "I take all the calves (from a neighbour) after about the beginning of March. Every Wednesday I send for what are above ten days old, as up to that time they require their mothers' milk, which is unfit for the dairy. The price I pay is 30s. each. They have for the first three or four days two or three quarts of milk at a meal; then gradually some food in the shape of gruel is added, and by degrees water is substituted for milk. Mixing oilcake with gruel is the secret of success. I use half oilcake, the best I can buy. Take a large bucket, capable of holding six gallons; put into it two gallons of scalding water; then add 7lb. of linseed cake, finely ground, which is obtained by collecting the dust that falls through the screen of the crusher, and passing it through one of Turner's mills; well stir the oilcake and water together, and add two gallons of hay tea. The hay tea is made every morning by filling a small tub with sweet hay, pouring on scalding water; use this in the evening; add a sufficient quantity of scalding water to the hay leaves, and cover down for next morning. The hay tea is very sweet, dark in colour, and I think the extract from the different herbs assists digestion. Again the mess is stirred, and 7lb. of mixed flour well worked in; the mixture consists of one-third wheat, one-third barley and one-third beans; add sufficient cold water to fill the six gallon bucket, and well stir. Two quarts of this with two quarts of cold water will be sufficient for a calf at a meal, and about the right temperature. The food should be given at regular hours—say six in the morning and six at night. Each bucket of gruel will be a meal for twelve or fifteen calves, and costs about 1s. 6d., or 3d. a day for each calf. We always measure the food with a two quart cup, and never overload the stomach of young calves. After fifteen days, when the calf chews the cud, some of the difficulty and danger is passed, and when the calf eats well we gradually diminish the gruel." The calves are tied up whilst being served, and Mr. Ruck prefers the old-fashioned plan of letting them suck through the cowman's fingers, as this prevents bolting, and a proper quantity of air is taken in, which assists digestion. As soon as they can eat, crushed corn, sweet hay and roots are placed within reach; vetches as soon as ready, and mangolds, of which a supply should always be stored if practicable. The calves live in a cool well-ventilated house, are kept very clean and quiet, supplied with fresh water daily, and the manure frequently removed. The addition of the decoction of hay is a very sensible practice, supplying some, if not nearly all, the nutriment of good hay, which the calf cannot otherwise obtain. Since the object of the feeder should be to imitate nature as closely as possible, we would suggest the introduction of a small quantity of sugar, just sufficient to give the requisite sweetness of new milk. Sugar plays an important part in the juvenile economy, and we find it present to a large extent in milk. It will also take the place and act the part of the cream which the calf gets in the natural state, although this may be more directly imitated by using a small quantity of palm-nut meal, a substance containing over 20 per cent. of fatty matter identical in composition with the fatty matter of cream. We must introduce this substance by degrees, for animals do not take it very kindly at first, on account of its gritty texture; but if used in conjunction with sugar, we are satisfied from our own experience that it would do good. Mr. Ruck advocates the use of a little hay cut up with straw, especially when the calf is just deprived of its liquid food. During the first winter the following mixture is recommended: 5cwt. of straw chaff, 5s.; 10cwt. of pulped mangolds, 5s.; 1cwt. of oil cake, 10s.; and 4cwt. of mixed crushed corn at 30s.; put together and allowed to heat moderately. This gives a ton of stuff superior to hay for 50s.; a small quantity of hay might, however, be added with advantage. Mr. Ruck touches upon the diseases of calves, referring especially to murrain, husk, scour, and lice, and gives some simple rules which are familiar to stock breeders. Thus murrain, which is a stoppage in the circulation of blood in the extremities, is easily distinguished by the crackling under the skin when the hand is passed over the infected part. It usually occurs when calves that have been badly kept are suddenly put into a luxuriant though watery pasture; blood is made too fast, and the system is unable to get rid of the excess of carbon. Prevention is better than cure, which is seldom possible; therefore always keep stock thriving. With regard to husk, Mr. Ruck believes it arises from a threadlike worm in the windpipe, resulting from ova taken in with lattermath grass, and therefore recommends that calves should, if turned out, be always kept on land that has been fed in the spring. The risk of disease from this cause would be obviated from yard feeding. Scour, which is very common with badly fed calves, may be prevented by generous diet regularly supplied. Of course, it is impossible always to have our animals in health, but nine-tenths of the maladies of young stock arise from mismanagement. Everything depends upon close attention to details, feeding with great regularity, supplying the proper amount of food to each animal in a suitable condition for rapid digestion, taking care that young stock are well housed, yet allowed air and liberty, and keeping the pores of the skin open. Another and not insignificant argument for good treatment is found in the greater strength thereby provided to withstand sudden attacks of disease, and rally from their effects.

CHAPTER III.

PRINCIPLES OF FEEDING—NATURE AND VALUE OF DIFFERENT KINDS OF FOOD.

E now proceed to consider the question of feeding, which may be divided into several heads; thus we may settle the age at which an animal can be brought out, the nature and value of different kinds of food, and the profit or loss that may be looked for from the operation. Evidently our views upon the first point will depend upon the nature of external circumstances. In the case of very rich feeding land, such as the Leicestershire pastures, the best return will generally result from purchasing well-bred full-grown animals in fresh condition, and finishing them off rapidly. Such land is too valuable to do the earlier work, which may be effected on cheaper material. Wherever sufficient straw is grown to provide litter and some forage, it is often on such farms a good plan to purchase in the autumn, winter well on straw and oilcake, and thus turn out fresh in May. At any rate, it is good policy to secure a portion of stock before winter, and make up our quantity in the spring, buying more or less according to the prospects of keep and the state of trade. On such land, breeding and rearing seldom pays. A cow, living on three acres, would not do as well as three fatting animals, from each of which we may look for £5 or £6 as the return for grass consumed. Summer feeding may be dismissed without further comment, save to remark that we should always try and purchase well-bred animals, such, as a rule, having greater tendency to fatten; that some sort of shelter should be provided during the heat of summer; and that the use of a very moderate quantity of cake is often found, especially during the latter stages of feeding, a very profitable addition to natural food. Cake is most commonly used, because the most easily given, and less liable to waste than meal; a mixture of oil-cake and crushed beans would often be an improvement, especially in wet weather or when the grass grows fast. At such times the grass is laxative, and the linseed cake, instead of correcting, rather increases this tendency. Well-ground cotton-cake might be substituted for linseed cake; for, being obtainable at a reduced price, and being equally feeding, it is, weight for weight, much cheaper food. There are some who altogether condemn the practice of outdoor grazing as extravagant, and tending to a loss of food; and it must be confessed that it is not always easy to regulate the mouths to the growth, and have the food constantly eaten to the greatest advantage. If the keep increases on the beasts, then it is trodden down, soiled on, and, so far as these animals are concerned, wasted; but after the last lot goes to the butcher, and the land has lain three or four weeks to sweeten, our comparatively hungry store cattle gladly gnaw up all that is left, so that very little is wasted. The injury from flies is often very considerable, especially when there is no shade or shelter; but this may to a great extent be prevented. No doubt animals tied up in cool, airy sheds, and supplied with grass *ad libitum*, would, on the whole, fatten faster than animals at large; but the injury to the quality of the food by mowing would be so great as to render the plan unfeasible. Lastly, we might save the trampling of the grass by a system of tethering, which is carried out on some dairy farms with Alderney stock; but our beasts would fret at first, and suffer much from the fly, and, we fear, do but indifferently well. We have long been convinced that the most paying practice, when the nature of the grass will allow it, is to give cake or its equivalent on grass during summer, and supply the market with moderate weights when meat is most scarce —viz., during the summer months. To do this, the animals must be brought forward during the winter and early spring with generous food. They should be placed in well-sheltered yards, to which a roomy shed is attached, and fed with a mixture of pulped roots, chaff, and meal—indeed, much in the same way as if feeding, only the nourishing food in smaller quantities. Those who have good grass and a large proportion, and whose arable land, both from its nature and limited proportions, is not capable of supplying roots to a large extent, will act wisely in not attempting winter feeding, but content themselves with growing and developing the cattle destined for summer grazing, which can be done without roots at all, although a few pulped and distributed through the chaff help the latter down wonderfully. Winter feeding is an expensive necessity for those who have but little grass, and whose land requires liberal supplies of fold-yard manure. In such cases the problem to be solved is, how to obtain the maximum return at the least cost. Few pretend to say that house feeding can be made to pay *per se*; but great will be the advantage if the increase in the animals covers the outlay in food, and we have the manure as our profit, for that represents a very considerable item. We may calculate that during four months (which is about the average time fresh beasts require to be housed) each animal will make from ten to twelve yards of manure, which at 6s. a yard—a fair price for such manures— gives a return of from 60s. to 72s. per head. Now, we believe that this result can be attained by careful management, and we proceed to sketch out a programme which we have adopted for some years, and which has proved satisfactory. The forwardest animals are drawn out from the rest, at the period when the lattermath is ready for pasturing. They are supplied, by means of cattle cribs, with a mixture of equal parts of decorticated cotton cake and palm-nut meal, two pounds of each per head, given in the mornings. At first they do not eat this well, picking out the cotton cake, but a few days suffice to bring them to their food, and they soon let one know if it is not given punctually. Not only does such a mixture supply a large proportion of feeding material, but the rather binding nature of the cotton cake tends to correct the too laxative influence of the grass. According to the state of the weather, these cattle are brought into the yards about the middle of October, and have pulped turnips, chaff composed of a mixture of hay and straw, and 6lb. of meal, i.e., 2lb. of barley meal added to their previous food. The proportion of pulp depends upon our resources; about 70lb. to 80lb. per day is amply sufficient, which is about half the quantity that would be necessary if the animals were fed on the old plan. The reason for placing them in yards is to bring them by degrees to a life of confinement; if at once placed in the boxes or tied up, they sweat so much as to loose flesh. They should remain in the yard about three weeks, after which they will settle in the boxes. Many complain with justice of the expense of boxes, which is quite double that of byres; but it must be allowed that animals thrive much faster in them than when tied up, and the manure is most excellent. We do not believe in stale fermented or cooked food; hence the pulp is made each day, and consumed within twenty-four hours. The animals are fed three times a day with the mixture, and the cotton cake being in lumps is given and readily eaten by itself at noon; the meal, that is, the palm-nut and the barley, being scattered over the pulp and chaff. Rock salt should always be supplied, and the animals require water; it is best when this is laid on, and always at command, but, as this is often not possible, we supply it once a day. Cattle eating 80lb. of roots with 40lb. of chaff will drink from four to six gallons a day. Some people recommend the use of the brush and currycomb, and no doubt the circulation improves when the pores are kept open; but in boxes cattle can rest themselves and keep themselves clean better than when tied up—moreover, the expense is considerable. If cattle are tied up in byres, we believe it would aid the process if they were turned out daily to stretch their legs. At first they would probably run about a good deal. Young growing animals especially require a little exercise, and any loss of force would be made up by improved appetite. We have found animals feed quite as rapidly in a sheltered yard, with plenty of shed and crib room, as when tied up. When within about a month of the market, we add 2lb. of the best linseed cake, making eight pounds a day of artificial food. This is expensive feeding, as we cannot calculate such food at much less than 1d. a lb. We calculate the cost of house feeding as follows:—$\frac{3}{4}$ of a cwt. of roots, 4½d.; 6lb. of artificial food, say at 5d.; hay and straw used as chaff, say 20lb. a day, at £3 a ton, 6d.; attendance 2d.; total, 1s. 5½d. per day, or 10s. 2½d. per week, charging a full price for the hay and straw. It will be found good work—more, we suspect, than an average result—if the animals increase 14lb. a week of dead weight; and as the market value even at present rates hardly reaches the cost, it will be evident that stall feeding can seldom be made to pay expenses and leave the manure clear. The farmer must consider himself well off if he gets his manure at half the price it would cost him to buy; and such manure as will result from box feeding on the food described is well worth the price we have estimated, viz., 6s. a ton. Feeding cattle is a necessity on light thin soils. On strong land artificial manure and thorough cultivation will produce remunerative crops, at any rate, for a number of years; but we cannot do without fold-yard manure on weak sandy soils.

Where we rear and feed—which with good management will on mixed farms pay, as a rule, better than buying in animals to feed—there can be no doubt that we should feed from birth, never allow the animal to stand still, much less go back, and thus bring our animal out at an early period. How this can be done will depend upon circumstances. Our practice in this respect has undergone a wonderful revolution in the last ten or twelve years. By pursuing the system sketched out, it is quite possible to bring out two-year-old steers averaging from nine to ten score a quarter, and equally practicable to have our heifers producing their first calves at the same age. Unless our soil is rich, and the produce very nourishing, we may not be able to make beef so economically at two as at three years. In the latter case we force less, but always should have the animal gently thriving. Where practicable, we recommend turning the animals off at two and a half years old, as a quick return is a great point in these days; but this requires a continuation of good feeding during the second year. If in the field, the pasture must be abundant, and night and morning trough food must be given. Comparatively early in autumn the beast should be housed. A well-ventilated box, which allows of more freedom of motion than the stall, is preferable for young beasts. The taste of turnips should have been previously acquired by a few roots scattered about the pasture. With the chaff and corn they are familiar enough, and so we get them to settle down after a day or two without a stand-still period, so common to sudden changes of diet. During the second summer at grass we recommend a mixture of bean meal, oil-cake, or cotton-cake, and barley meal, always given with chaff, the quantity of artificial being increased from 2lb. at the spring to 4lb. in the autumn; and it may be as well to pulp a few turnips in August to add to the mixture, and thus more thoroughly accustom the animals to their winter food. For a few days after first coming in, a portion of green food might be given, so that the change should be as gradual as possible. By such means we cannot fail, under careful superintendence, to keep our stock thriving, and by May or June, when our roots are on the wane, we may expect some very pretty ripe beasts, weighing from 90 to 100 stones of 8lb. dead weight. In the event of a third year being considered advisable, we should winter in yards the second year precisely as during the first, taking care to run them thinly, and feed moderately well. We have used green German rape with good effect in such cases, breaking it up and steaming, or else macerating it in boiling water, and then pouring it over the chaff. The price of this article has considerably risen of late years; moreover, it is difficult to obtain, and we should consider palm-nut meal as equally cheap and more feeding, on account of its greater richness in fatty matter. This, however, brings us to another part of our subject, viz., the nature and value of different kinds of food.

The system of feeding horned stock, though more rational than of old, is still in many districts costly and extravagant, and must be amended if we hope to make it a profitable operation; or, perhaps it would be more correct to say, escape a heavy loss. In former times, the high price of wheat rendered it desirable that rich manure should be made, even though the expense of its manufacture was great; especially was this the case at a time when the farmer could not, as now, supplement his home-made manure by artificials; consequently we find that large quantities of costly food were given, the greater part of which remains in the manure. We know that even now, in some cases as much as 10lb. to 14lb. of linseed cake are given daily, a third of which only is assimilated. Now, such treatment is wasteful and unscientific. We cannot afford to keep animals merely as expensive machines for the manufacture of manure; and unless it be possible so to feed cattle as to cover our expenses, we had better abandon the practice altogether, and obtain our manure from other sources. We believe that it is possible so to feed as to make both ends meet. Let us examine the progress that has been made. In Scotland, where feeding was largely pursued in the days before linseed cake was known, the custom was to give sliced turnips *ad libitum*, and long hay. The quality of the swedes being good, excellent beef was made, though the process was rather slow, and decidedly expensive. This barbarous practice is still the rule, though we are glad to know that the pulper is slowly forcing its way into favour. In vain has Dr. Lyon Playfair shown that heat is an equivalent for food, and that every manifestation of force is accompanied by loss. Scotchmen, as a rule, still persist in pouring gallons of cold water, in the form of watery turnips, into their bullocks, thereby lowering the temperature of the body to such an extent that the animals may be seen to shiver after a hearty meal, and much extra fuel (*i.e.* food) is consumed in raising the temperature thus needlessly lowered. A large ox will eat as much as from $1\frac{1}{2}$ to 2cwt. of sliced roots a day; 90 per cent. being water, we have more than twenty gallons of fluid: how much heat must be absorbed in raising this to the temperature of the animal's body! The effect is similar to pumping too much cold water into the boiler of an engine. The temperature falls rapidly, and extra fuel is required to regain the original condition. By the introduction of linseed-cake a reduction was effected in the quantity of roots and hay, but the system of feeding was still defective. The mixture of different materials into a compound suitable for digestion was rendered possible only when the chaff-cutter, and more lately the pulper, were introduced, and we must regard these inventions as of the utmost importance to agricultural success. Every farmer who keeps stock, whether he fattens them out or only grows them, should have a chaff-cutter and pulping machine. The mere fact of reducing the food into a form which renders it more easily digested, though important, does not half express the value of these inventions. It is the means they afford of economising the more costly part of our food—viz., the roots—and enabling us to substitute cheap straw for expensive hay, that gives them such importance, and marks their introduction as inaugurating a new era in cattle management.

Pulping-machines are constructed on two distinct principles, viz., first, such as have the cutters fixed on a barrel, differing only from Gardener's turnip slicers in the size of the blades and the spaces in the barrel, through which the cut food escapes; and, secondly, such as have the knives fixed in a vertical disc. The latter have important advantages, and principally that the roots are not rolled round and round, and bruised, the form of the hopper being such as to allow of their remaining stationary whilst being cut; whereas when the root comes in contact with the revolving barrel, there is a tendency to fly off, and thus a rolling motion is communicated, the root is bruised, and the juice more or less extracted, which is to be avoided. Of all disc pulpers yet brought out, that of Messrs. Hornsby and Sons stands first; not for the quantity of work done in a given time, but for the perfect way in which the roots are brought into the required condition with the minimum loss of juice. The difference in this respect in different machines is really remarkable. At the Oxford trials in 1870 samples of the pulp were taken and examined a few hours after being cut; the difference in colour and freshness was very great; in some cases, notably the produce of barrel cutters, the pulp of mangold was already nearly black. The disc cutters cut to the last piece without the root being squeezed. The price of a pulper is so inconsiderable, the largest power size being only £6 10s., and the saving is so great that, as we said before, it is only ignorance of its value that prevents universal use. In order to have the greatest advantage from pulp, it should be fresh. Advocates for fermentation are not wanting, but we have always noticed that if kept to the third day the cattle do not eat it so well. Whether in case of necessity the pulp might not be compressed and kept from the air, especially if slightly salted, we cannot say, but think it very probable and worth testing. With steam power, which is only at command one or two days a week, the preservation of the pulp becomes an important question; but with horse-power always at hand we like using the pulp fresh, never keeping it more than from twenty-four to thirty hours. The improved one or two horse gear is particularly adapted for driving chaff-cutter and pulper. We are acquainted with a farmer who has about 100 head of cattle of all ages in his boxes, stalls, and yards, during the winter. A strong Galloway pony does the work, the operations of pulping and chaff-cutting being performed separately; the chaff, as cut, falls down on to the floor where the pulper stands. The largest feeding beasts get 80lb. a day; the younger animals, which are not feeding, but growing, from 20lb. to 40lb. The store animals have more straw chaff than the fatting beasts, the rule being to allow them as much as they can eat up clean. The mixture of roots and chaff offers an excellent medium for the distribution of artificial food in the form of meal.

Animals thus fed should have water offered to them once a day; the cattle in the yards can drink when they like. The fatting beasts will usually take about two or three gallons a day. Bearing in mind Liebig's views as to the loss sustained by each motion of the body, there must be a decided gain in presenting the food in a state requiring so little labour in mastication. The work of filling the belly is effected in a much shorter time than formerly, consequently there is more time for rest, which is a condition favourable to the deposition of fat. The number of animals we can feed out is usually determined by the supply of roots. If, then, by pulping, we can economise the roots by at least one-third, it follows that the system allows of extra stock to that amount being made out. Taking the time of feeding to be six months, and the quantity of pulped roots to be 80lb. against 120lb. sliced, it follows that we shall save fully 3 tons, worth at least 30s. a head. Winter feeding of cattle is often held to be an expensive necessity, and no wonder, with the costly method too frequently employed. We believe that, under judicious management, the winter feeding of cattle is the cheapest method of maintaining and increasing the fertility of our land.

We have said that in many districts the system of feeding is still costly and extravagant. This is very apparent in the wasteful practice with regard to straw. Now, if properly made and taken due care of, a large portion of the straw should be used as food, as it bears a considerably higher value for this purpose than as mere manure. The quality varies considerably, according to nature of soil, climate, and, above all, method of harvesting, the difference in this latter respect being very remarkable and worthy of most serious attention. Some years since we experimented on a crop of white Poland oats, cutting one portion very unripe, another green but still ripening, and leaving a third portion until dead ripe. We may dismiss the first, as the analysis proved that material was left in the straw which should have gone to the grain. The second—cut green, but sufficiently forward to allow of a full and fine sample—contained as much as 10 per cent. of sugar and gum, 30 per cent. of digestible fibre, and $2\frac{1}{2}$ per cent. of soluble (*i.e.* available) protein compounds. In the over-ripe the per-centage of sugar, gum, &c., was reduced to 3·19, the digestible fibre to 27·75, whilst indigestible woody fibre, useless matter, was increased from 31·78 to 41·82 per cent., the soluble protein being reduced to 1·29. The effect of ripening straw is in all cases to increase the insoluble material at the expense of that which is available as food, and therefore the practical question for each to arrive at is the exact period of time when the greatest amount of nutriment remains in the straw, and the grain is properly matured. It will be seen, by a careful study of the analysis of these straws, made by Dr. Vœlcker, and recorded in the twenty-second volume of the "Royal Agricultural Society's Journal," that the actual amount of solid matter is identical in the two samples; therefore the grain is not robbed, and moreover early cutting secures a finer quality, and is economical in many respects. Now let us compare this oat straw with a sample of well-made clover hay. We shall find that the difference of nutritive matter is not so considerable as we might suppose—

	Clover Hay.	Oat Straw.
Water	20·50	16·00
Oil	3·59	1·05
Soluble protein compound	5·00	2·62
Insoluble ditto	8·75	1·46
Sugar, gum, &c.	13·07	10·57
Digestible fibre	16·42	30·17
Indigestible fibre	25·62	31·78
Soluble mineral matter	4·43	3·64
Insoluble ditto	2·62	2·71

The hay, though much richer in oil and protein compounds, does not contain so large a proportion of non-nitrogenous elements as the straw, and we may fairly calculate that the straw is worth two-thirds as much as the hay. Now hay, such as this, would be valued at about 50s. a ton for consumption, and therefore the straw may be estimated at from 30s. to 35s., whereas the actual manuring value of the same would probably not exceed 10s. to 12s. The loss from treading such material into manure, instead of passing it through the animals, must be apparent. Let us remember, moreover, that other straw is even more valuable than the oat. Pea haulm, cut early and well harvested, is most valuable, quite equal to much of the artificial hay from poor land, and such material should be taken the utmost care of. This fact is not recognised, and in valuations between tenants we find that pea haulm is put in some counties actually lower than wheat straw.

Bean straw, especially if the pods are attached, is more valuable than is generally supposed; and though, from its hard woody nature, it is not so adapted for food as softer straws, still a portion may be used with advantage; while as we are now in the habit of cutting it in a much greener condition than formerly, it should not be overlooked as a source of food. Bean straw is increased in value by steaming, principally by being made softer, and therefore more easy of digestion, but to a limited extent its solubility is increased. Enough has been advanced to show that straw is a valuable part of our natural food, and when properly made may be economically substituted for hay in the feeding of cattle. The cost of chaff-cutting, on the most economical plan, will not exceed about 6s. a ton. Now, the use of chaff enables us to effect a great saving in the consumption of roots; and as roots hauled off the land are very expensive food, costing probably from 7s. 6d. to 10s. a ton, this is one great feature of modern farming. The pulper, by mashing up the roots, causes the juices to mix with and saturate the chaff, and thus we produce a composition that is readily eaten, and obtain the same effects as formerly with half the quantity of roots. A fatting animal does not require more than about 70lb. to 80lb. of roots, which may be distributed through some 20lb. of straw.

On the use of artificial food, in which we include home-grown grain as well as purchased materials, we find a great diversity of practice, partly arising from varying conditions, and partly from ignorance of the digestive processes. We shall not arrive at uniformity until the feeding of the cattle has been made the subject of carefully conducted experiments. The commercial mind, accustomed to subject every process to experiment and test, will be surprised to find that questions of such importance as the exact proportions of natural and artificial food most suitable at different ages and stages of feeding, are still matters only of opinion, but there are great difficulties in the way of demonstration. The comparison between living subjects is always influenced by the variation that exists in constitution, temperament, and feeding quality; and it is a nice point, and opens a large field for error, to allow for these disturbing causes. It would be important to ascertain exactly what proportions of natural and artificial food prove most economical at different ages and stages of feeding. We are quite certain that money is often lost from over-feeding with rich nitrogenous foods, especially in the early stages of the process. The proportion which the system can take in and convert into flesh is so small, that much the larger part goes into the manure; and, remembering the vast complicated stomach of the ox, it is clear that a bulky food is required, which contains a considerable portion of indigestible fibre. The proportion between the stomach and intestines affords an indication as to the nature of the food required by different animals. Thus for each 100lb. live weight the ox has 11¼lb. of stomach, and only 2¼lb. of intestines; the sheep has much less stomach and more intestines, and altogether a smaller percentage of digestive apparatus, indicating the necessity for more concentrated food; whilst the pig has only 1½lb. of stomach and 6lb. of intestines to each 100lb. live weight, demonstrating that concentrated and generous food is required. These are important points, indicating the kind of food most suitable to each. Thus bullocks have been fattened entirely on good straw and oil cake. Such a food would not fatten sheep, as their digestive apparatus needs a considerable proportion of starchy food; whilst pigs do best when fed entirely on meal. In the absence of direct experiments on the subject, we venture to suggest a dietary suitable for animals growing and feeding at the same time, and which would be applicable to the animal from the period of weaning until death, the quantity to be given depending on age. It is well to remember that there is virtue in a mixture of different substances, even though we might supply the same constituents in a simple form. Variety suits the digestion: linseed cake, or cotton cake, according to market, 2lb.; barley, or wheat meal, or palm-nut meal, or, better still, an equal mixture of all three, 4lb.; beans, peas, or lentils, according to market, 2lb.; locust beans or malt, 1lb. The whole being reduced to a fine powder, and thoroughly mixed. In the above we have a due admixture of the flesh-forming and fat-producing elements. During the later stages of fattening we might discontinue the beans, and increase the proportion of either linseed cake or palm-nut meal. Commencing with the calf when weaned, ¼lb. of such a mixture daily distributed over the chaff and pulped roots would be ample, to be increased to 1lb. for the first winter, 2lb. during the second; and when the animal is put up for feeding, 4lb. to commence, and 7lb. to 8lb. to finish, will be found quite sufficient with a due proportion of roots and chaff. If we are anxious to overfeed animals—as for show, for example—it is necessary to stimulate the appetite by the use of tonics and carminatives, just as the East Indian with a damaged liver and bad digestion craves for pickles; but the use of such food in early life must be injurious, and cannot be recommended.

We need hardly insist upon the advantage of using articles good of their kind, well grown, properly matured, and free from adulteration; though more costly, they are in their effects cheaper than inferior articles. Linseed oil-cake ranks at the head of all our purchased foods, being from its complex nature admirably suited to feeding purposes; the chief bar to its use is the high price of a genuine article, and the great difficulty of finding such. The proportion of really pure cake made is small compared with the quantity of inferior, and that, again, differs according to the variety of ingredients of which it is composed. In Yorkshire this has been so much felt, that at Driffield the farmers have made a company for crushing pure linseed, and thus have secured a first-rate article. Manufacturers have been blamed for the low quality of their cake, when too often the fault rests with the consumer, who is not willing to pay the higher price. Great progress has been made towards sound views on this matter, and cheap rubbish is not now, as formerly, eagerly bought up. The publicity given by the Royal Agricultural Society to adulterated and inferior makes has been most serviceable. It will be a great misfortune to the farming community if, from prudential consideration, the Society cease to pursue the course they have adopted, and which has done more to gain them respect than any other of their useful efforts. The quality of a cake may be ascertained to some extent by a careful examination under a microscope, and by maceration in hot water; but it is wise before making a large purchase to have an opinion from a good authority.

We may here pause to notice the reasons why linseed cake possesses such high feeding qualities and is so generally esteemed. In order to arrive at a satisfactory answer, we must first inquire what the animal system requires. First and foremost, matter that will supply the daily waste of muscle, or which will build up new muscular structure; secondly, that which provides fuel for keeping up the heat of the body, and which when supplied in excess of these requirements can be deposited as fat; and, lastly, mineral substances suitable for the construction of bone. The following analysis of Dr. Voelcker may be relied on as giving the contents of a good average cake.

Moisture	12·44
Oil	12·79
Nitrogenized, or flesh forming principles	27·28
Heat-giving substances	41·36
Mineral matters (ash)	6·13
	100·00

The above requires but little consideration. Linseed oil cake contains a large proportion of two most valuable ingredients—ready-made oil and ready-made flesh. Mr. Lawes has proved by experiments that vegetable oil is two and a half times more valuable than the class of starch compounds which exist to so large an extent in most vegetable substances: hence one explanation of the value of linseed cake. The oil being so valuable, the use of the seed has been advocated. Indeed, we believe that the system of feeding animals in boxes was introduced by Mr. Warnes, in order to feed on linseed. There are two good reasons against the use of the seed. First, the value of the oil for commercial purposes is too great to allow of its economical use; and secondly we may have too much of a good thing. When properly mixed with other food, linseed is very feeding, but its action in excess is purgative; and cheaper material can be found, which is equally effective. We can well understand the value of linseed cake for animals eating nothing but hard dry straw; thousands of beasts are so wintered, and thrive. Linseed cake has special advantages for growing animals, owing to the large proportion of ready-made flesh formed. We may supply fattening food from other sources, at less cost, but we question if any food save decorticated cotton cake can be found so valuable for growing stock. With such qualities, it is not surprising to find a lively demand, and consequently a price which is excessive in comparison with other feeding materials. Mr. Hope, late of Fenton Barns, as illustrating the change that had taken place in the comparative value of corn and horn, mentions, at a discussion before the Edinburgh Chamber of Agriculture, that forty-five years since, when a child, he remembered one of his father's men coming from Leith with two carts of linseed cake, describe how the people of Preston Pans came out of their houses wondering what he had in his carts, and that he had told them it was a kind of "bannocks," but he did not think it would suit those who drank tea.

When linseed cake is over £10 to £11 a ton, we question whether it may not be replaced with advantage by cotton-seed cake, green German rape, or palm-nut meal. The first of these is now largely used, and, when free from coarse shell, is a safe food with a considerable amount of feeding property; rich in nitrogenous elements, it is best when mixed with such a material as palm-nut meal, which contains a high per-centage of fatty matter: the prices of these articles are at least 50 per cent. lower than linseed cake. We would especially caution our readers against the use of any but the best samples of cotton-seed cake, viz., those that are yellow in colour, and finely ground. Several deaths have been clearly traceable to coarsely ground cotton-cake, the indigestible fibre of which has accumulated in the intestines, and caused inflammation. Twenty years ago this article had no existence; then the cotton-seed was thrown on one side as useless. At first the cake was made from the whole seed, afterwards machinery was invented for removing the black outer skin, and as this formed a considerable proportion of the whole, and consisted entirely of indigestible fibre, the superior value of the decorticated cake, as a feeding material, was soon established, and has since maintained and increased its position. Dr. Vœlcker wrote a valuable paper in the Royal Agricultural Society's Journal for 1868, from which the following figures, being the mean of seven analyses, are taken:—

Decorticated Cotton Seed Cake.

Water	9·28
Oil	16·05
* Albuminous compounds (flesh-forming matters)	41·25
Gums, mucilage, sugars, &c. (heat-producing materials)	16·45
Indigestible fibre	8·92
Mineral matter (ash)	8·05
	100·00

* Containing nitrogen 6·58 per cent.

If we compare the above with the analysis of linseed cake, we shall find the cotton-cake superior in the proportion of oil and flesh formers. The former is not like linseed, purgative in its action, but is sweet and agreeable to the taste. Cattle eat it readily. Dr. Vœlcker considers it quite equal to linseed cake. We think it very superior to much of the rubbish sold under the title of linseed cake, although it can be bought at one-third less price. The cake we are describing is the thin cake, imported chiefly from New Orleans, and which finds its way into Liverpool. It should be of a bright yellow colour, close texture, and, when cut with a knife, the surface exhibits a shining appearance, caused by the oil. Only occasional specks of black should be visible. Any larger portion of shell indicates imperfect decortication. This cake is admirably adapted for cattle on grass, as its slightly astringent property tends to prevent scouring, which the luxuriance of the grass would otherwise produce. The large percentage of flesh formers renders it peculiarly valuable for growing animals. Decorticated cake is generally about a quarter less than the best linseed cake, and for many purposes we consider it decidedly superior. It must not be forgotten that Mr. Lawes places it at the head of his list as a manure producer, and we cannot imagine any more immediate or certain way of restoring fertility to land than by applying manure made from cotton-seed cake.

Very much the larger portion of the cotton-cake used in this country is undecorticated, the English make nearly entirely so. When made from clean seed, it is a wholesome food, although much less nutritious than the decorticated cake. The following is an analysis by Dr. Vœlcker:—

Water	10·53
Oil	6·10
* Albuminous compounds (flesh forming)	22·62
Gums, mucilage, &c.	26·48
Indigestible fibre	26·96
Mineral matters	7·31
	100·00

* Containing nitrogen 3·62 per cent.

A comparison of the two analyses proves that decorticated cake is nearly twice as valuable, although seldom costing above one-third more. Hence, at present rates, we think it decidedly the cheaper food. Another point in its favour is being so much more digestible. Common cotton-cake cannot be safely used for young stock. It is likely to cause indigestion and stoppage. Good cake should be yellow, but much darker and with much black shell. It is, however, easily distinguished from a very worthless article made by pressing the refuse from decorticated cake with a small portion of common seed. A good deal of such rubbish has been sold, and several fatal cases have followed its use. Dr. Vœlcker describes the death of a bullock belonging to a Mr. John Fryer, of Chatteris, which was fed upon such cakes with mangolds, barley-meal, and clover hay. The post mortem showed the paunch enormously distended with food, the lower stomach quite empty, the duodenum for twenty-four inches in length entirely blocked up with two or more pounds of the irregular-shaped concave and comminuted husks, which were found to be identical with the husks in the cake. No wonder such was the result, for the cake contained more than half its weight of husks. A slight acquaintance with the appearance of cotton-cake will enable the farmer to distinguish good, bad, and indifferent. Although we have never suffered from the use of the common cake, further than with calves—for which it is not suitable—we strongly advise our readers to buy the best decorticated. It must be remembered that cotton-cake is a binding, rather than a relaxing food, hence it is not suitable for animals feeding entirely on dry food. Under such circumstances the addition of linseed cake is desirable, but as a mixture for growing stock that get turnips, and especially for cattle on grass, we have a very high opinion of the decorticated cotton-cake.

Rape cake has not acquired the importance it deserves, for two reasons: animals do not eat it readily, owing to its hot bitter taste, though when once accustomed, there is no difficulty on this score; secondly, inferior samples, especially English made, frequently contain the seeds of wild mustard, the oil of which is a violent poison. Fatal cases have occurred from this cause. The oil, being volatile, is easily destroyed by subjecting the cake to a certain temperature: hence if we are at all in doubt as to the quality of the cake, boiling or steaming is a safeguard. The presence of mustard can be readily detected by treating a portion of finely powdered cake with boiling water. The oil, which is very volatile, escapes, and may be recognized by its pungent odour and powerful irritating action, making the operator's eyes water. According to Dr. Vœlcker's analysis, good rape cake should contain—

Moisture	10·68
Oil	11·10
Nitrogenized, or flesh-forming matters	29·53
Heat-giving substances	40·90
Mineral matter (ash)	7·79
	100·00

It is evident from the above that rape cake is not deficient in nutritive properties, and Mr. T. Horsfall states, in his valuable "Articles on Dairy Husbandry," that he used it with marked success. His mixture for milch cows comprised, rape cake, 5lb., and bran, 2lb., for each cow, mixed with bean straw, oat straw, and shells of oats, in equal proportions, supplying this food three times a day *ad libitum*. The materials were moistened, well steamed, and given in a warm state. Mr. Horsfall further states that he had no difficulty in inducing his animals to eat the cake, and never found the butter affected by it. We have known others equally successful. Few people steam now, even when there are facilities. The practice is not on the increase, and we must continue to regard the hot, unpleasant taste of rape cake as a serious impediment to its extensive use.

NATURE AND VALUE OF FOOD.

Smith's *Palm-nut Meal*, which results from the grinding and pressing of palm kernels, possesses considerable fattening proporties, owing to the large percentage of vegetable oil which it contains. This oil is very similar in appearance to lard, and appears to be readily convertible into animal fat. Dr. Vœlcker's analysis is as follows:—

Moisture	5·92
Oil and fatty matter	20·01
* Albuminous compounds (flesh-forming matters)	13·87
Mucilage, gum, sugar, &c.	38·24
Woody fibre (cellulose)	18·56
Mineral matter (ash)	3·40
	100·00

* Containing nitrogen 2·22 per cent.

We have connected the meal with the name of Messrs. A. M. Smith and Co., of Kent Street Oil Mill, Liverpool, the original importers of palm kernls, because we believe they are the only crushers who sell a meal with this high percentage of fatty matter. Others subject the meal to a second pressure, and thereby reduce the oil to 12 or 14 per cent.; whilst by a chemical process employed on the continent it is possible to dissolve out nearly all the oil. As it is perfectly certain that the value of this food depends upon the oil, it is evident that the sample containing 20 per cent. must be much more valuable than that having only two-thirds that quantity. The meal is dry and harsh looking, palm oil being solid at ordinary temperatures. It is sweet, and will keep good for any length of time. Some patience is required in accustoming animals to its rather gritty taste; when once they take to it, however, they eat it freely. For cattle on grass, when liable to scour, we consider its use in conjunction with decorticated cotton-cake is of great value; the mixture possesses high feeding properties, and affects the manure much in the same degree as linseed cake. The price of this meal is from 7*l.* to 8*l.* a ton.

Indian corn or maize is an important feeding material, and when either soaked or ground into meal, forms a useful mixture with other substances. It is largely grown in the United States, and, with improved railway communication in the future, we may anticipate increased supplies. We have not been able to find a detailed analysis, and therefore quote from Mr. Horsfall's comparison of different foods for dairy cows. He gives—

Oil	7	per cent.
Starch, sugar, &c.	60	,,
Nitrogen	2·25	,,
Mineral matters (phosphoric ash, ·19; potash, ·17)	·36	,,

The albuminous compounds required to furnish 2·25 of nitrogen would amount to nearly 14 per cent., which leaves about 18 per cent. for water and indigestible matters. The large proportion of starchy matters, the moderate percentage of flesh formers, and the deficiency in minerals, all indicate that Indian corn is more adapted for feeding than growing animals, also that it is not a food to be used alone. Thanks to Mr. Lawes, this is not a mere speculative opinion. In his pig-feeding experiments, the results of using Indian corn alone are given in the following table:—

No. of Pigs.	1st Period of 14 Days.	2nd Ditto.	3rd Ditto.	4th Ditto.	Total Period of 8 Weeks.
1	31lb.	6lb.	40lb.	19lb.	96lb.
2	15lb.	13lb.	13lb.	13lb.	54lb.
3	12lb.	17lb.	19lb.	23lb.	71lb.
	58lb.	36lb.	72lb.	55lb.	221lb.

The following, in explanation of the above, is extracted from the report which will be found in the fourteenth volume of the "Royal Agricultural Society's Journal," page 472:—"One of the pigs gained more than 2lb. a day during the first fortnight of the experiment, but the other two only about half as much. Before the end of the first period it was observed, however, that this fast-gaining pig, and one of the others (No. 3), had large swellings on the side of their necks, and that at the same time their breathing had become much laboured. It was obvious that the Indian corn meal was in some way defective diet, and it occured to us that it was comparatively poor both in nitrogen and mineral matter, though we were inclined to suspect that it was a deficiency of the latter rather than of the former, that was the cause of the ill effects produced. We accordingly determined to continue the food as before, but at least to try the effect of putting before the pigs a trough of some mineral substances, of which they could take if they were disposed. The mixture which we prepared was as follows: 20lb. of finely-sifted coal ashes, 4lb. of common salt, and 1lb. of superphosphate of lime. A trough, containing this mineral mixture, was put into the pen at the commencement of the second period, and the pigs soon began to lick it with evident relish. From this time the swellings or tumours, as well as the difficulty of breathing, which probably arose from the swellings, began to diminish rapidly. Indeed, at the end of the second period, the swellings were very much reduced; and at the end of the third they had disappeared entirely. Notwithstanding this serious drawback, it was found that the animals were satisfied with less of this food, though so poor in nitrogen, in proportion to their weight, than, with one exception, of any of the others; and it will be found that the increase is satisfactory when compared with the food consumed." Indian corn, at any rate for pigs, possesses considerable feeding properties, and there is no reason to doubt its value for cattle when judiciously mixed with other food. As a general rule, we may consider it cheap when it rules 4s. to 5s. a quarter below grinding barleys; up to £8 a ton it is reasonable; latterly, owing to reduced supplies, the prices have been unusually high.

Barley is largely used for feeding. Either the coarser samples of home-growth, which are not suitable for malting, or the hinder ends—that is the tail corn which comes out of finer samples—is available for meal, and when well harvested, has a marked effect in the production of flesh. Foreign barley makes excellent meal, being generally drier and harder. The advocates for the remission of the malt tax would have us to believe that better results would be obtained on malt. Mr. Lawes, however, is of a different opinion, and as his conclusions are the results of direct experiments, we place most confidence in them. The increase in weight of sheep on a certain quantity of barley was considerably greater than on the same after being subjected to the malting process; and he says, "Not only is the weight of the malt considerably less than that of the barley from which it was produced, but that weight for weight, independently of loss and cost of process (estimated at 2s. per quarter), the feeding qualities of malt are not superior to barley. At the same time, he admits that, as a mixture with other food, or as an occasional stimulant to digestion, malt may be usefully employed. The composition of barley is as follows.—

Water...	14
Flesh-formers	14
Starch, &c.	68
Fatty matter	2
Ash	2
	100

Beans, peas, and lentils are so identical in composition, and similar in their effect, that they may be substituted for each other according to our convenience, and may be considered together. These are valuable feeding materials, and, when used with judgment, give satisfactory results. Owing to mechanical condition, and also to the large proportion of flesh-forming element, all three are more or less indigestible if given without due preparation, are partially wasted, and if largely used, are apt to cause constipation of the bowels. Beans, which are the hardest, should either be broken small or ground into meal. Peas are much softer in their nature, and will be sufficiently prepared by being kibbled or broken small. Lentils, which in their natural state are, owing to a hard skin, very indigestible, should be reduced to a coarse powder. The predominating feature of these substances is the large percentage of flesh-forming material, which points out their peculiar value for working horses, growing animals, or for young stock that are growing and feeding at the same time. We give Dr. Cameron's analysis:—

	Common Beans.	Peas.	Lentils.
Water	13·0	14·0	13·0
Flesh formers	25·5	23·5	24·0
Fat formers	48·5	50·0	50·5
Woody fibre	10·0	10·0	10·0
Mineral matter	3·0	2·5	2.5
	100·0	100·0	100·0

Bean meal is admirably adapted for calves after they are weaned, and when, either out at grass, or receiving green food in the yards. Peas appear particularly suitable for young sheep. Lentils may be substituted for either, provided they are properly prepared. Egyptian beans are largely imported, and, weight for weight, are, when sound and good, quite as valuable as home-grown corn; indeed, from their drier condition, we should prefer them to new home-grown beans, which, owing to the presence of more water, are not always desirable food.

Locust, or carob bean, when ground, forms a considerable percentage of some of the condimental food, and might with advantage be used in small proportions, as it contains a large quantity of sugar; but, like German rape cake, the supply is uncertain, and always very limited. Either this or malt is desirable to the extent of 10 or 15 per cent., as tending to render the food more palatable, but neither is to be used in large quantities. Sugar is too soluble for the ruminant, and much saccharine food tends to cloy the appetite. Such at least has been our own experience, and we cannot coincide with much that has been written as to the use of malt. As an alterative, especially for sheep that are out of health, we believe it will prove most valuable; but as a constant food in large quantities we do not think it desirable.

In addition to the above, and especially in the latter stages of feeding, something in the way of a condiment may be given, as encouraging an appetite that has become a little delicate. The following mixture has been used with success: Fenugreek seed, 32lb. at $1\frac{1}{2}d.$; mustard, 8lb. at $2d.$; linseed, 8lb. at $1\frac{1}{2}d.$; carraways, 4lb. at $4d.$; fennel, 4lb. at $5\frac{3}{4}d.$, making a total cost of $8s.$ $11d.$ for the half-hundredweight, or in round numbers $2d.$ a pound. From 2oz. to 4oz. a day would be sufficient for each animal. The expense is trifling, and the effects very satisfactory.

We come now to the question of profit or loss—and here much will depend upon our skill in buying well, selecting such animals as have aptitude to feed, and not paying more than they are worth. The cost of feeding may be variously estimated, according as we place a higher or lower value on the materials we use. Taking the roots at $10s.$ a ton, and straw at $1l.$ a ton, and artificials at $8l.$ a ton, the cost per week will be as follows:—5cwt. of roots at $6d.$ a cwt., 140lb. of straw, at $1s.$ a cwt., 42lb. of artificial at $8l.$ a ton—total, $8s.$ $2d.$ We have to estimate on the other side the progress and the value of the manure. We think it may be fairly assumed that a beast on such food as above should gain 12lb. of dead weight weekly:—12lb. of meat at $6\frac{1}{2}d.$, value of manure, say at $2s.$ a week—total, $8s.$ $6d.$

Nothing but experiments carefully carried out can give us data on these points. We believe that under judicious management fatting may be made, not perhaps to pay in one sense, but to be the means of providing a large supply of powerful manure which it would cost a good deal of money to replace in artificial manures.

CHAPTER IV.

BUILDINGS, AND THE MANUFACTURE OF MANURE.

THE importance of providing shelter for our live stock is generally acknowledged; the economy of preserving the animal from the depressing influences of a low temperature and exposure to rainfall is no longer questioned. It is not now, as formerly, necessary to explain how food supplies combustible material wherewith the temperature of the body is maintained, or to show that the quantity of such food will bear a direct proportion to the temperature in which the animal lives, and that an equable and rather high temperature is the most favourable for economising the food. Nor is it necessary to dwell upon the injury arising to the health and growth of the animal when, in place of being comfortably housed, it has to face the pitiless blast on a bare pasture, or to stand in an open yard, over fetlocks in sludge, as we too frequently see in our dairy districts. These questions, we repeat, are now fully understood, and it is rather with a view to moderate opinions, so that we may not exaggerate the value of shelter, that we propose to discuss the question of covered *versus* open yards.

It is always difficult to get a new idea entertained, however good it may be, and a long time must elapse and much discussion is required before the agricultural mind is prepared to receive any novelty; but when once the crust of prejudice is broken through, and we are fairly afloat, there is the danger of being carried away by the tide that sets in, and landed far beyond the point we aimed at in starting. Enthusiasts, with more ardour than judgment, are prepared to advocate their hobby under all circumstances. It appears to us that the question of covered yards is in danger of being injured by the intemperate zeal of its advocates. We believe that, under many conditions, the covering of our homesteads will prove an economical investment. The saving of straw and concentration of manure are points of great importance. But, on the other hand, we are perfectly certain that manure may be made in open yards, when properly constructed, without any loss of valuable materials. "How is this?" we hear some one exclaim. Has it not been stated by Alderman Mechi and others that manure made under cover is worth much more, weight for weight, than that from open yards? And have we not all heard of the losses to which manure is liable when exposed to the action of rain? The graphic picture of the hillside yard, deluged with water collected by the long unspouted roofs of stable and barn on the upper side, with the duck-pond below, which receives the filtrate, has been too often brought forward as the type of the open yard system. But is this fair? As well might we condemn covered yards because failures have occurred through improper construction, causing imperfect ventilation. Divested of exaggeration, we believe the real facts of the case are these. Manure may be made in open yards properly constructed, without the loss of valuable materials; but a greatly increased quantity of litter will be required, and must be supplied according to the greater or less rainfall, in order to soak up the liquid and prevent waste.

This is the most cogent argument against open yards. We cannot, in many cases, afford to throw away so much straw, which, if properly harvested, has a considerable consuming value; and although not going so far as some who advocate its entire use as food, we believe that the perfection of manure-making consists in avoiding all excess of straw, and using no more than is absolutely required to absorb the solid and liquid excrements. Manure in open yards varies in quality in its different layers, according to the state of the weather when those layers were formed. If dry, little or no fresh litter will be required, whereas in a wet time litter must be spread once or perhaps twice a day. The proportion of excrement, therefore, differs greatly; and in order to obtain a uniform bulk, we must mix the different layers in a heap and occasionally turn, in doing which we are liable to some loss of valuable ingredients. Then, again, we have extra labour in carting and spreading; serious items, if we consider that thirteen loads of covered dung are equal in effect to twenty loads from open yards. An open yard should have a watertight floor, either quite level, or sloping slightly towards the centre, be of moderate size, capable of holding some six or eight beasts, and the hovel properly spouted. Under such conditions, if we are careful about the litter, good though bulky manure may be made; and this is consolatory for those who for many reasons may be obliged to adhere to the old system.

It is evident, then, from what has been advanced, that the amount of rainfall materially influences this question; that it will be much easier to make good manure on the eastern side of England, where the rainfall does not exceed twenty inches, than in the west, where we have frequently double that quantity: consequently, we should expect to find covered yards most in use on the western side of our island. This, however, is not so, as, with one or two exceptions, we find there is a total absence of such arrangements. Now, why is this? Some may say *this* is an evidence of the superior intelligence and enterprise of eastern farmers, and that after a time their example will be followed. To such we would remark that in Norfolk covered yards have existed for many years, and in these railway days farmers are very soon alive to the improvements in other districts; we think, therefore, the more natural conclusion would be that for some reason the covered yard does not suit the circumstances of the case. Now what are these reasons? We are going to hazard a rather strong assertion.

We believe *that covered yards are not adapted for breeding animals*, and that as breeding is pursued generally in the districts we have named, therefore the system of covered yards does not find favour, and may probably never be used to any great extent. If this difficulty could be overcome—as we fear it cannot—then these very farms, often consisting of two-thirds pasture, present the exact conditions under which the greatest advantage would follow the adoption of covered yards. In such cases, straw is always a scarce article. To economise the straw, and at the same time make our manure to the greatest advantage are objects for which much might be sacrificed; but when it comes to a question of the health and constitution of the young animal, we naturally pause. We hear some one exclaim, "Ventilate." Well, we agree with our imaginary friend that this is absolutely essential; but it will not remove the difficulty. Take a number of yearling heifers in the autumn, divide them into two equal lots, place one in a covered yard, and let the others occupy a yard and shed, and perhaps enjoy the range of a dry grass field. Supply them with the same kind and quantities of food. The housed animals may look the sleekest, carry most flesh, and possibly not eat quite so much food daily; but at May-day take the two lots into a fair and offer them for sale. The rougher coats will command considerably the most money. And this is no idle prejudice, as we have proved by experience. Housed animals, when first turned out, often lose ground instead of improving, and sometimes never thrive all the summer. The truth is, the open yard is the more natural system, better calculated to develope bone and muscle, and to inure the constitution to climatic changes. Under cover we have conditions that favour the deposition of fatty matter, the sinews are inclined to be soft for want of proper exercise, and, though growth is often rapid, the flesh that is made wants compactness. It may be said that it would be easy to arrange for our stock being exercised. We might drive them about for a time daily; but this could not be compared to the voluntary motions of an animal, with the free air of heaven circulating about him. It is all very well for professors to talk about the waste of food from a cold atmosphere, or the absorption of an increased quantity of oxygen during exercise, and to dilate upon the advantages of warmth. We listen to such arguments with the greatest deference when applied to fattening; but we know that exercise and fresh air, aye, and rain too, are all beneficial and necessary agents in developing the frame and flesh, and giving vigour to the constitution of our breeding animals.

If it were possible to keep our animals entirely under cover from birth to death, then we should advocate the system, and on arable farms where breeding is attempted, the produce being destined for the butcher at the earliest possible age, this might be successfully carried out, and great weights attained at an early period. Again, when animals are bought in for winter grazing, the use of boxes, or even tying up under cover, is an excellent system, provided we have good ventilation. The farmers in the eastern counties, with farms chiefly arable, require stock in large quantities to feed. They produce a great bulk of straw, which must be made into manure, either passed through or put under the animal. Animals placed in roomy compartments, with plenty of air, at an even temperature, must thrive faster and be in a more healthy condition than when crowded up in a feeding house, with little head room, and often exposed to draughts. This is almost self evident. Neither do we agree with those who see an objection to covered yards, on the score that when straw is abundant, it cannot be consumed. Such an argument might have had weight in olden times; but now, with the pulper and chaff-cutter, a considerable portion may be advantageously passed through the animal. And, indeed, this economising of straw we regard as the greatest argument in favour of covered yards. If cattle cannot be purchased, sheep will feed admirably under cover, provided due care is paid to their feet.

In constructing new premises the question of covering the yards should receive weighty consideration, as it may be done at much less cost during construction than afterwards. Mr. Bailey Denton, in an excellent paper read before the Central Farmers' Club some years since, estimates that the cost will not exceed 6d. a square foot covered, when arranged for in the architect's plan; whereas, if a roof is put on existing premises, the cost, without calculating for external walls, which most likely are there already, will vary from 8d. to 1s. per foot. As each beast requires at least 100 feet, it follows that the outlay per head is from 3l. 10s. to 5l. Mr. Moscrop, in a clever article in the Journal of the Royal Agricultural Society, gives the cost, from actual contract, at about 3l. per 100 feet. A friend in Norfolk, who was one of the first to cover a yard, informs us that, valuing everything—timber sawing, masonry, &c.—the cost was about 3l. 10s. per 100 feet covered. We may therefore fairly assume that, according to circumstances, the cost of covering old buildings will vary from 3l. to 5l. for every beast we can accommodate. 6l. per cent. on outlay will clear off the principal in thirty years, and thus, taking the highest cost we have the sum of 6s. to add to the other expenses of each feeding beast. Is this money saved in the superior condition of the fattening animal, in the greater economy of labour in moving and manufacturing the manure, and the value of the straw per head which remains for feeding purposes, but would, under the old system, have been trodden underfoot? We answer, without hesitation, in the affirmative; and moreover believe that a handsome balance remains over in favour of covered yards for *fattening* beasts.

The want of proper accommodation for cattle is very apparent, in many of our principal grazing districts, cattle being treated much in the same way that they were a century and more ago. As a rule, the management is bad, or rather to speak more accurately, there is a total want of any in many cases, the animals being left out in the grass fields all the year round. The want of proper accommodation is often the cause of this, and the landlord is more to blame than the tenant. Without good buildings, properly fitted up, both the farmer and his stock are more or less at the mercy of the elements. The buildings on most grazing farms are quite inadequate to the wants of the cattle they are intended to protect. We believe that by a little trouble and a small outlay they might be made much more comfortable than they are generally found. No one accustomed to travel about the country can have helped seeing the mess a lot of store cattle make in a field, especially near the shed and gates during the winter months. These sheds, which are so common in grass fields, and which are intended to protect a lot of beasts in winter from the inclemency of the weather, may be made much more useful at little cost. They now consist simply of a shed with manger and rack, a food house in one corner, and perhaps a small rickyard at the back; in few instances do we find a yard attached, consequently the animals cannot in the wettest weather be prevented from poaching the land, thereby rendering its surface more like a ploughed field than anything else. We would propose to have yards made to these sheds and a gate attached, so that the animals can be shut in; if the yard be walled in so much the better, from its being so much warmer than if merely railed round. By a little contrivance the shed may be made available for two or even four fields.

The yard need not be large, and must be made in proportion to the number of animals that the shed will accommodate, so as to give them all room without the stronger ones knocking their weaker brethren about too much. Unless they are tied up this cannot be helped, the underlings getting less food than the others.

We strongly advocate the plan of tying animals up at night and when they are feeding. A considerable saving of space will be thus ensured, for, strange as it may at first appear, a shed and yard will accommodate more animals if tied up than if they be left to range about. There are several other points to be urged very strongly in favour of having all sheds made so that the animals can be tied up, besides the economy of space. The equal distribution of the food is of great importance, all faring alike. When the animals are loose the principle of might being right holds good, and the weaker ones necessarily go to the wall. The effects of a winter spent among a lot of stronger beasts will be seen long after turning-out time; and it is doubtful if it be ever got over, however well the animal may be tended afterwards.

On most farms where sheds are built out in the fields for the accommodation of cattle, straw will be found to be very scarce. There are few farms where there is not some, but what there is ought on no account, if properly harvested, to be used as litter. And now comes the question, What are we to use instead? Sand has been recommended very strongly by Mr. Brereton, in his paper on "Stocking Land," in the "Royal Agricultural Society's Journal." There may, however, in many cases be very great difficulty experienced in getting sand, and, even if attainable, the price may be too great to make it of service. The same may be said of sawdust, which will answer the purpose equally well. We are not now speaking of littering the yard, but the shed where the animals are tied up at night. The amount of litter required will of course depend upon the floor of the shed and the fall from it. If the fall be imperfect, and it lies wet, more litter will be required; but if a proper fall be attained the animals ought to require little or none. Sand and sawdust are merely useful in so far as they render the surface more readily cleaned, and not for their actual properties as litter. Irregularities in the surface of the floor will always hold moisture, and otherwise conduce greatly to the discomfort of the animals. Care should therefore be taken to have the floor as smooth as possible, and to provide sufficient fall to carry away all the liquid manure. This manure need not be wasted, but it is not our intention now to discuss how any waste may be avoided.

The more impervious the floor the better, and for this purpose we recommend asphalte; a good foundation of rough stones or some similar material being laid first, and then the asphalte on top of it. As some of our readers may be ignorant how asphalte may be cheaply made, we will attempt to explain its manufacture. Take either gravel stones or broken flints of about half an inch in diameter, and have the dirt carefully sifted out; then spread them on a wooden board with side ledges, and pour gas tar on a little at a time, turning over the stones or flints with a shovel until every stone is wet with tar. They cannot be turned too much. Care must be taken that no tar is left besides what the stones take up. The next process is to add enough sharp sand to make the mixture of the consistence of thick pudding, turning it over as before. The floor on which the mixture is to be laid should be carefully levelled, and then it should be laid on from three to six inches thick. A little sand must be sprinkled over it to prevent the roller sticking to it, and then roll with a heavy roller. This can be done during the summer months, when the shed is not required, and when there will be every chance of the surface getting quite firm and hard before the cattle begin to tread it. It must on no account be used for a month, and should be rolled as often as possible until perfectly hard.

Before dismissing the subject of tying up cattle in what would otherwise be open sheds, we would remark that no divisions will be requisite between the animals beyond the posts to which the chains are attached. If wooden divisions be put up they would interfere with the utility of the shed if it should be required as an open one at any time; whereas, if there are no divisions, it will be almost in the same state it was in before being altered.

And now we come to the mangers and racks. The latter we believe to be a most useless appendage, and one that we expect to see every year less common. The manger, if properly made, will answer the purpose of a rack, if long hay or straw be given, without distressing the animal in getting its food. Cows are not intended by nature to eat their food as a giraffe does, and therefore they cannot with ease to themselves pull hay out of a rack placed on a level with their horns when standing upright. Every animal should have a separate manger, or rather division of manger, to itself. For this purpose, upright partitions must be placed in it opposite the posts to which the beasts are tied, and again in the middle between them. In order to prevent the chaff from being thrown out and wasted, iron rods must be driven through from front to back; three of these rods will be requisite, in order to prevent any waste at all for each division of the manger. Unless this precaution is taken the cattle will be found to toss their food out with their noses, and especially will this be the case if there be any pulped roots in it. In searching for the pulp among the chaff they waste the whole mass. The mangers should be at least from twelve to fifteen inches deep, and should be of uniform depth, back and front, and will be best on the ground, or nearly so; just high enough to insure sweetness and freedom from waste.

The yards, if they are already in existence, will be generally found to lie very wet from having been hollowed out somewhat every time the dung was removed. To obviate this evil we would recommend their being filled up to a proper level with burnt soil ashes—the scouring of ditches, tussocks, and, indeed, all refuse, properly burnt will answer the purpose. If the cattle are allowed the range of the yard during the day, it surely is of moment that they should have a dry instead of a wet bed, and if we do not tie up the animals at night we may be sure that, however dry and warm the shed may be, the yard will still be the resting place of the underlings, and ought to be made as comfortable as possible.

When a plentiful supply of ashes can be secured the yard may be cleaned out to the bottom every year, or, indeed, oftener, but where they are difficult to procure we can see no reason why they should not remain down until enough can be burnt to supply their place. These ashes make an invaluable manure for root crops, and, indeed for any crop, and are not expensive in manufacture. On mixed arable and grass farms a plentiful supply can always be secured, and on purely grazing ones, if care be taken, enough can be got for the purpose. The cost of burning will of course depend upon the ease with which the materials can be got together and upon the nature of the soil—strong clay or calcareous soils burning the best, sandy soil being very difficult to burn. In some cases coal or wood will be required to help the fires. These ashes, if placed in a yard, should be laid so as to give a fall on all sides, and will require a small quantity of straw on them to prevent the cattle sinking during wet weather.

We are no advocates for costly premises; whilst advising that good durable materials be employed, we would have the thing that was necessary at the least cost, consistent with durability. Bricks and slates and foreign timber will supply our wants in most cases, but we may be able to wall with rough stone; or, in districts where coals are dear, we may be driven to use a substitute for bricks If the soil is a retentive clay, for instance, we can use clay-lumps with advantage—the process is very simple. The clay is spread on the ground and thoroughly puddled by horse-treading; short straw, couch grass, or any fibrous vegetable matter being added in sufficient quantities to insure cohesiveness. When sufficiently tempered, the clay is made into lumps in an ordinary mould, and stacked up to dry, the stack being protected from rain by thatched hurdles. In this way it soon dries and becomes fit for use. The lumps of clay are fastened together with a mortar made of clay, and known as pug; and, provided the eaves are made to overhang such buildings, are sufficiently durable. In chalk counties we have another substitute for bricks: the lower beds are best for this purpose. Though quite soft when first quarried, exposure for a few months causes the surface to become hard, and if protected from the action of moisture by well-spouted overhanging roofs, such walls are permanent, and when jointed with black mortar, have a pleasing effect. Lastly, where gravel abounds, we may make use of concrete, which is probably the most durable material, next to granite, that can be found. Moreover, the process is so simple, and the knowledge of how to prepare the materials so easily acquired, that ordinary labourers can execute the work, under the supervision of a carpenter, and the cost is not more than half that of bricks. We commend Mr. Tall's lecture, delivered before the Society of Architects, as worthy of careful perusal by all who are about to erect farm premises.

It is a great mistake to pinch ourselves for room, especially in buildings intended for live stock. Several evils arise. The health of the animals suffer from want of sufficient air, and their comfort is interfered with from the confined position in which they lie. The feeding alleys and dunging passages are so narrow that necessary work is either imperfectly done, or takes up more time than would be the case if the space was larger. It should always be remembered that economy in operations that have to be frequently performed, is more important than in those which are only occasional; indeed this consideration should influence the arrangement of the buildings in reference to each other. Ventilation, however ably managed, cannot overcome the evils arising from insufficient area. The change of air, instead of being gradual, as it should be, is so rapid as to cause draught, which is more or less unhealthy.

In designing buildings for the accommodation of cattle, regard should be had to the method of feeding: if it is proposed—as assuredly is desirable—that pulped roots and chaff be used, then a good root-house, with chaff chamber over, must be placed in such a position, and with such access to the principal yards and feeding-sheds, as shall ensure the least expenditure of labour in feeding. A building, 18ft. square, or 18ft. by 20ft., 9ft. high to floor, and 6ft. above, making the walls 15ft. in all, will be sufficient for a farm of from 200 acres to 300 acres. The root-house must have double doors, large enough to allow of a load of roots being backed in and shot up, and the position of the pulper and the entrances so arranged as to allow of the maximum quantity of roots being stored. A sufficiently large pitch door in the chop chamber must be provided, so that both hay and straw can be readily delivered from a cart. Of course it would be convenient if the roof-space of adjoining buildings allowed of the storage of straw, but such a provision entails great additional cost, and there are other objections to the plan, such as increased risk of fire, and the possibility of the fodder being injured by the moisture arising from the animals below. We prefer that the materials should be carted as required: a horse and cart, and one man to assist the feeder, is all that is requisite. The hay would be cut from the stack, and the straw come from the Dutch barn. Both pulper and chaff machine can be driven by a horse gear, and if the operations are performed singly, one horse-power will suffice. Even where steam power exists, we should advise the horse machinery because the pulp must be made daily, and it would not answer to get up steam for such work. An old horse should be told off for this work, and with carting the roots, getting fodder, and working in the gear, the time will be pretty well occupied.

The particular arrangement of the buildings will depend upon circumstances, which it is impossible to define, except we have the case before us. The feeding-house should join the root-house on one side. Supposing we tie up our cattle, which is the cheapest way of housing them; the house should be 18ft. wide, walls about 7ft. 6in. or 8ft. from floor, feeding passage communicating with root-house, 3ft. 6in.; manger, 2ft.; length of beast space from back line of manger, 6ft.; width of gutter, 12in.; passage way behind beasts, 5ft. 6in.; stalls for two animals 7ft. in the clear; one or more doors, according to the length of the house, opening into the fold yard, for the removal of manure. For growing stock, open yards with wide shelter sheds, are desirable; the sheds not less than 15ft. inside measurement, and provided with a brick manger, so that the cattle can be fed under cover, and if necessary, tied up at night or at feeding time. These yards should not be too large; ten or twelve beasts are sufficient for a yard. A yard, 40ft. square, with a shed 40ft. by 15ft. would secure ample accommodation for ten or twelve two-year-olds. It can easily be arranged that the shelter sheds should form one side of the yard, easily reached from the root-house. Two of such yards, at least, will be required. And the cow-house, if dairying is pursued, may be placed on the opposite side of the root-house, and form part of a second yard. Supposing that we adopt the system of box-feeding, which has much to recommend it, the most economical arrangement is a double row of boxes, each box 9ft. square, with a feeding passage in the centre 4ft.; thus requiring a building 22ft. wide in the clear.

The animals should be separated by posts and rails, which may be made to remove. This will prove convenient when the manure is taken out. If practicable, water should be laid on. Animals eating a full supply of sliced roots will not drink; but if fed on pulp, and especially if the weather be hot, fluid is beneficial. The water troughs, like the mangers, should be movable, so as to be easily raised, as the manure accumulates, and should be fed from a pipe above. It will be found desirable to have a round rail to slide up and down above the manger, which can be raised by the beast when feeding, but which protects the manger at other times, and prevents what otherwise is unavoidable, viz., the frequent presence of the animal's fæces in the manger.

The height of the building is important in connection with proper ventilation; many an otherwise good feeding house has been spoilt from being too low. We think if the walls are 6ft. above ground, and the building open to the roof, ventilation may be satisfactorily arranged for by leaving spaces in the walls under the eaves, providing wooden shutters so as to regulate the admission of air according to temperature, and by raising every fourth ridge tile, making it overlap, thus allowing for the escape of heated air, whilst rain and snow are effectually kept out. This is so simple and inexpensive, that we are surprised to see the plan so seldom carried out. The roof, whether of slate or tiles, should be rain-proof. The joints between the tiles must be carefully pointed. And here we may diverge from our subject to remark that the addition of gas tar to ordinary mortar immensely increases its adhering properties. Half a gallon of common gas tar to a bushel of lime makes a cement which resists the strongest wind, and such a mixture is highly suitable for bedding ridge tiles, or pointing inside or out.

The buildings thus described would occupy the centre of the plan, proceeding from the root-house and opening upon each of the yards or courts, so that the manure could be removed, and the litter supplied; this could be effected by means of slide doors, one door answering for two boxes. Thus, then, a simple and economical arrangement of buildings is suggested: cattle boxes down the centre, on either side of the root-house open sheds, the third side of the yards being made up with stables, piggeries, and cow-house. It is impossible, without detailed drawings, to give further particulars, but it will be evident to our readers that we look upon farming as a manufacturing of meat, and must arrange our buildings so as to contain the requisite machinery, and raw material so placed as to secure the maximum result with the greatest economy. The arrangements for the preparation of corn for market are quite of secondary importance; all the space formerly devoted to the storing of sheaf corn can be dispensed with. Dutch barns and portable machinery effect the object in a much simpler manner, and a granary over the cart lodge should be useful, not for storing the corn, but for dressing it for market.

And now as regards the comparative advantages of boxes or stalls, for feeding animals, we think it must be clear that the loose box, if properly ventilated, will prove the most comfortable to the animal, as offering less contrast to life out of doors than the stall. The young animal has room to move about, lie down, and rise up in any position, and freedom to rub—a point of considerable importance. When once settled, beasts thrive better in boxes than in stalls; and young animals especially, which grow their flesh as well as lay on fat, must have freedom to move about. The box system tends to a cheerful, contented state; the beast can see and even lick his neighbour, if inclined; there is no hindrance, though possibly *a bar*, to intercourse; and, thus being in company, the beasts rest better than if quite apart. The minimum size for a box should be 9ft. by 9ft., the depth of the box from the surface not to exceed 2ft. Formerly it was customary to have the pits 3ft.; but this is a mistake, as it is inconvenient for getting the beast in and the manure out; and, with care as to the supply of litter, we shall find that the depth named will hold the manufacture of three months. Nine feet square will be sufficiently large for ordinary beasts; but we should prefer 10ft. by 9ft., giving the extra foot in width on account of the space occupied by the manger. The great objection to the box is the cost. Two animals can be stalled in the same space. This, however, may be partially remedied by having a double row and gangway common to both. If young animals are stall-fed, they should be allowed to stretch their legs once a day, or, if this is not practicable, the skin should be regularly brushed over daily with a strong brush. The irritation encourages circulation and exhalation, and the healthy state of the skin has much to do with progress. What a sanitary lesson cattle teach us, as they stand for hours exposed to rain, though a warm and dry shed offers a tempting lair. Nature points out the importance of this *al fresco* bath; the rain causes reaction and increased circulation; the pores are kept open. When housed, nature's flesh brush is lost, and we must try and compensate. In the box, they can rub themselves in every direction, and so do well.

Instances may occur in which it is desirable to feed all the year round. Mr. Mechi and others adopted this plan as the only means of making sufficient manure; but it cannot be recommended with any idea of profit. Our animals live far more cheaply out, either in the grass fields or in yards, and generally come in about October. They should be gradually accustomed to the change; a yard and shed for two or three weeks will be an intermediate arrangement that must be beneficial. Great attention must be paid to ventilation; the cattle house cannot well be too cool at first. The sudden change induces sweating, and we have known cattle brought at once from pasture into boxes lose more weight by sweating during the first two months than they gained by food. We think in the case of very strong coats the scissors might be used with advantage; certainly it is very desirable to check excessive sweating. If first placed in yards at nights, then kept altogether in yards, and so quieted down and accustomed to change of food and manipulation, a considerable step will be gained, and progress in the boxes will be rapid and without check. Indeed, we do not think a beast should be in the box longer than suffices to fill it with manure, and then, walking out fat on the level, he makes room for a successor. Therefore the preliminary attention in the yards is most important, bringing them on through the first stages; so the feeder secures a succession of animals, picking out as a start the forwardest, and working on through his lot.

Inasmuch as the principal object in cattle feeding is the manufacture of manure, it may not be considered out of place if we devote a few lines to its consideration, not only as to the actual making of manure, but also as to its preservation. Much misapprehension has and does prevail on the subject, illustrating the old proverb, that "a little knowledge is a dangerous thing." Agricultural teachers of early days have something to answer for, since they dwelt upon the volatility of ammonia until the puzzled farmer believed that every smell proceeding from his muck heap meant serious loss of valuable elements. We remember when this was such a well-established point, that an ingenious theory was devised to reconcile the good effects of manure spread out and exposed for weeks on clover leys, in preparation for wheat. According to this theory, the exposure of manure, especially in a decomposed condition, to the action of the air, would result in a frightful loss of ammonia; but the under surface of the clover leaves were believed to possess a power of absorbing the ammonia, and so collecting it ere it could be lost in the air. Such ideas have long since been dissipated under the searching lights of modern science. Dr. Vœlcker studied the practice of successful farmers, and experimented to ascertain the scientific explanations, which he presented to the world in some admirable papers published in the "Journal of the Royal Agricultural Society." It is a consolation to know on so good an authority that, with ordinary care and management, manure may be manufactured and preserved without serious or even sensible loss, and this without the necessity for a heavy outlay in covering over yards or making boxes. Let us not be misunderstood. There are advantages in making manure under cover; prominently, being able to regulate the supply of moisture to suit the requirements of the case; having a uniform proportion of litter and fæcal matter, which results in a very even quality of manure; in the great saving—quite cent. per cent.—in the use of straw; and lastly, in avoiding loss from heavy gluts of rain, which, however carefully provided against, must result in some loss.

Were the manure the only question, we should strongly advocate the erection of covered yards, especially in situations where the rainfall is considerable, although we are aware that great care is required in supplying straw. If too much is employed, the manure is imperfectly trodden, fermentation commences, and, if not arrested, serious loss occurs from the manure becoming fire-fanged. The heat is injurious to the animals, and ammonia is dissipated in the process. Either the animals must be kept rather dirty, or else water must be pumped on the manure, so as to secure the requisite amount of moisture. The great objection, however, to covered yards is their unsuitability to young growing animals, which require exposure in order to gain hardihood.

We must therefore see how far manure can be economically made in open yards; and this will depend in great measure upon the construction of the yard. All surrounding buildings must be carefully spouted; the yard should not be large —40ft. by 50ft. is a good size, sufficient, with a proper shelter hovel, to accommodate ten beasts; the floor should be made water-tight, and should slightly incline towards the centre; the gradient should be very small; the yard as nearly level as possible; the walls and shed sufficiently high to allow of the manure accumulating without risk of the stock knocking off the tiles of the shed, or jumping the walls. Thus arranged, the absorption of liquid is a question of litter. There will of course be times when, owing to rainfall, a constant supply will be necessary, and must be used, otherwise the beasts will suffer or the liquor will escape, carrying with it a certain though small proportion only of valuable matter. The ingredients of fresh manure are neither very volatile nor soluble, consequently the high-coloured liquid which oozes from the fold-yard often looks more potent than it is; still, it should be retained, and when proper care is exercised, first in the form of the yards, and next in the distribution of litter, this can be done. When voided, excreta are not in a state of putrescence, and if properly incorporated with litter and sufficiently compressed, little or no fermentation takes place in the open yard. This is shown by the action of the air upon such manure when removed from the fold and placed in a heap. The fermentative process commences at once, and a few weeks effect a great change in the bulk and appearance of the manure. The manure is in a comparatively fresh state until exposed to the atmosphere. We are led to dwell on this subject in consequence of some incorrect views, as we think, promulgated in a lecture delivered to the Bakewell Farmers' Club by Mr. A. M'Dougall, who quotes Sprengel as to the supposed loss manure sustains in its manufacture. If urine is left exposed to atmospheric influence, no doubt putrefaction occurs, and volatile ammonia results; but if the urine is either collected in a tank, or, which is better, allowed to run into the yard and become absorbed by the litter, there is no loss whatever, for reasons that will be explained.

Whilst, therefore, strongly recommending the use of a disinfectant, such as the compound prepared by M'Dougall, and which comprises sulphurous and carbolic acids, to sprinkle the floors of cow-houses, pig-yards, &c., we cannot agree in the necessity for its use, either in the shelter sheds or open folds; although possibly such a compound might be valuable to dust into the manure heap when being turned. We have reason to believe that M'Dougall's Patent Disinfecting Powder is valuable in a sanitary and economical aspect, and no stock-keeper should neglect its use. All we maintain is, that foldyard manure not being in a putrescent state, such an agent is not required. Whilst on this subject, we may be pardoned for explaining the action of the disinfectant, which is a compound of two salts, viz., sulphate of magnesia and lime, and carbonate of lime, to which are added bone flour—*i.e.*, phosphate of lime—used to reduce the strength and add a valuable fertiliser. The gaseous products of decomposition are sulphuretted and phosphuretted hydrogen, either free or in conjunction with ammonia, though, as we shall see by-and-by, the latter is, when the decomposition occurs in connection with vegetable matter, provided for in another way. Sulphurous acid causes decomposition, being itself also destroyed, the resultant products being sulphur, phosphorus, and water. Sulphurous acid further acts as an anti-putrescent by its affinity for oxygen, absorbing that gas and preventing other bodies combining with it. Carbolic acid is the most powerful anti-putrescent known, and at once arrests fermentation in a remarkable manner. Its regular and frequent exhibition, not only on the floors and walls of the cow-houses, but by impregnating the air by wet cloths, &c., did more to keep off cattle-plague contagion than any other preventive measure, and there is strong evidence to show it was effective. The magnesia is introduced in order to form a double compound of phosphorus and ammonia, highly valuable as a fertiliser.

The great point necessary for the due preservation of foldyard manure is its compression; hence the advantage of small yards, into which the sweepings of the stable, cow byres, &c., are carefully spread. Now, for strong land, manure cannot be in too fresh a state, and therefore it may be removed at once from the yard and spread upon the land; but on light land, and in order that the manure may acquire a uniform condition, it is necessary that it first undergo a fermentative process in the heap; and during this stage loss may be incurred. A certain proportion of runnings, more or less, will break away from the heap; hence it is well to make the heap upon some good absorbent. The best we know of is red clay ashes, broken to powder; six inches of such a bottom is very valuable. The heap must be carefully built, the sides kept as plump as possible, and the top after a few days may be lightly covered with soil. Some moisture is necessary to start the process, but our soil prevents the washing effect of heavy rains. In a few days fermentation commences, and proceeds rapidly. Sometimes, in order to secure greater uniformity, the heap is turned, and during the process M'Dougall's powder might be freely scattered, as it would probably be useful in decomposing the compounds alluded to, and possibly saving a small quantity of ammonia—very little, however, as the bulk that has been set free in the decomposing process is already united with the products of vegetable decay, forming highly soluble but non-volatile compounds. These are liable to be washed away; hence the necessity for placing the manure on an absorbent, and the utility of the soil covering. It will be seen from what has been advanced that the danger is not from volatilisation, but from solution; therefore, the sooner decomposed manure is spread upon the surface, the better for its safety; once there, the drenching shower may work in elements which the grateful soil will give account of. The wind and sun can only remove moisture, and not injure.

In some situations, where the rainfall is considerable, and where litter is scarce, it will be found impossible to absorb the moisture as it falls, and in such cases a drain from the centre of the yard is required to convey the liquid into a tank, from whence it should be carted or pumped on to the land. Such liquid in reality contains very little manurial matter, because the ingredients of fresh manure are, as has been stated, not readily soluble. Still there must be a portion of the urine, and it is important that, whether more or less, such should not be wasted. We have no great faith in tanks or liquid-manure carts—the first are seldom watertight, and the use of the latter is frequently neglected; and therefore we prefer, where it is possible, the process of absorption by litter. The efficaciousness of the latter could be greatly increased if reduced into short lengths, as the absorbing surface is thus greatly increased. In the case of box-made manure, the value of litter, when in short lengths, is very much enhanced, and the expense of reduction will be repaid in the saving of straw and the more perfect character of the manure, which can often be cut out like cheese, and is, when thus made, exceedingly rich. With properly constructed yards, and due attention as to the distribution of litter, good manure may be made, and all the moisture absorbed—manure that, according to the proportion of fæcal matter it contains, will be found as valuable as that which is made under cover.

CHAPTER V.

DAIRY MANAGEMENT, THE MILK TRADE, &c.

OUR INTRODUCTORY CHAPTERS would be incomplete except we touched shortly upon Dairy Management. First, as to the method and cost of feeding. This will depend upon the quality of the soil, the proportion of arable and grass land, and the quantity and description of stock kept. On grass farms the cow lives entirely on grass and hay. If the land is rich, two and a half acres will be sufficient—viz., one acre for grazing, one and a half acres mown, and the lattermath fed. Many authorities consider that the hay from one acre should supply winter's food for five months; but it is not enough, as a full-grown cow, when not supplied with anything else, will consume two hundred weight a week, if allowed to do so. On poor land, four acres will scarcely suffice for summer and winter food; and the medium soils, which range between these extremes, will feed a cow on about three acres, always supposing that the cow grazes in the ordinary way. Probably the most economical method of consuming our food is to keep the cows tied up in a well-ventilated shed, and bring the food to them, mowing the grass as soon as it is sufficiently grown, and, in the case of arable land, depending upon successive crops of Italian rye-grass, vetches, &c. In this way we avoid that waste which is more or less inevitable when cattle seek their own food. Moreover, the cow is protected from flies, and lies cool and quiet; whilst in the fields they are made half wild by flies, which, in a woody country especially, are very trying; and this continual irritation powerfully effects the secretion of milk. The objection to this plan on grass land is that the quality of grass frequently cut soon degenerates. Another plan which has been tried with the Alderneys consists in tethering the cows, shifting them at stated intervals, sometimes as often as three or four times a day. By this system the grass is consumed very closely, and with economy. Heavier beasts, like the shorthorns or Herefords, will not do on this plan, as they could not get enough, and, moreover, would not eat clean.

So far we have only alluded to the old-fashioned plan of feeding with grass and hay; but of late years an artificial system has been generally adopted, and a considerable saving thereby effected. Animals are in some cases kept when there is no grass land. The late Mr. Hewitt Davis, writing to the *Agricultural Gazette*, describes a Surrey farmer who had been very successful with a dairy of thirty cows on a poor chalk farm of 210 acres, of which only five acres were pasture; the food in winter consisting of cut straw, bran, turnips, and mangold wurtzel, and in summer of tares, clover, &c. And yet upon this keep—so artificial and opposed to ordinary practice that many old dairymen would not believe profitable feeding could be carried out—the butter was admirable, and readily sold at remunerating prices.

In all cases where we have command of sewage on sandy soils, Italian rye-grass should be grown and cut from time to time for our cows tied up. Very heavy crops may be produced, and a great weight of cattle fed. The object of most dairymen is to have the calf born early in the spring, so that the cows may be ready to yield their goods when required. In such cases the cow should be allowed to dry, when she comes in about November, and during the winter she lives on straw, with a few roots and a modicum of cake, and is thus maintained inexpensively. When winter milk is required, the cow must live on a higher and more expensive diet. We proceed to describe what we have found to be economical feeding, commencing with that period, whenever it may be, that the cow is dried off in order to prepare for the next calf. With a few exceptions, cows should not be milked closer than within about two months of calving, as they then have time to get up condition, and the calf is always the stronger for this rest on the part of the mother. Occasionally we find cows that are such abundant milkers that they will yield freely to the last, and, indeed, require milking to keep them quiet. This peculiarity is, however, as rare as that of a cow we once knew of in Northamptonshire, which required milking three times a day. During the interval between drying off and the birth of the calf, the food should neither be rich nor too invigorating. If in winter, a run in a well-sheltered yard, or else a turn out on pasture during the day, and a supply of cut straw, with a few roots, night and morning; in summer, a short, rather poor pasture, will be most suitable. The cow, especially if of the Alderney breed, will require particular care as the time of parturition approaches; for, if mending too fast, the risk of uterine inflammation is considerable, and from such attacks they seldom recover. Some recommend taking a little blood as a precautionary measure, but we are averse to this as a barbarous expedient, and altogether wrong in principle, and greatly prefer to give a mild aperient in some warm gruel. For many years we adopted this plan, and gave, about three or four hours after calving, a dose consisting of from $\frac{3}{4}$lb. to 1lb. of salts, $1\frac{1}{2}$oz. to 2oz. sulphur, and $\frac{1}{4}$oz. of powdered ginger. In the case of very heavy milkers, and where the secretion commences some time before the cow is due to calve, it is a good plan to milk the cow once a day, commencing about ten days or a fortnight before.

Supposing our cow well over calving, and the calf removed after a week or ten days. The cow will now require good feeding, and on an arable or mixed farm we should recommend a variety of material, such as hay and straw chaff, with a small quantity of pulped roots, just sufficient to moisten the mass; some 12lb. or 15lb. daily would do well. Swedish turnips are generally objected to, as influencing the flavour of milk, but the quantity mentioned would hardly have any effect; at any rate it will be worth trying when roots are plentiful. We want in addition some cheap concentrated food or mixture of foods. Green German rape cake, when it can be bought, as is often the case, at 6*l*. a ton in London, is a good investment, as it does more for the milk than linseed cake. Palm-nut meal, especially rich in butter-making ingredients, has also proved itself valuable. A mixture of the two would not be amiss. Brewers' grains, when cheap, are principally valuable as removing the dry character of straw chaff. Steam, when available, might be made to pass through the mass; but we seriously doubt the economy of cooked food for a breeding animal, which depends upon its powers of digestion to extract nourishment from poor food. Bran is another fine milk food, though lately too expensive to answer. Supposing we adopt something of this sort: 2lb. of rape cake, 2lb. of palm-nut meal, and a double handful of brewers' grains, dispersed through a stone (14lb.) of well-cut straw chaff (a mixture of barley and oats would be best), with pulped roots. The cow will eat at three meals some 20lb. of dry food with as much more roots; and the cost of the whole at ordinary market and consuming value will not exceed 5*s*. a week, to which may be added 1*s*. for attendance. During the two months the cow is dry the cost will not exceed one-half the above. It will be seen that this is quite as cheap as a diet of hay only, even if we charge the latter at a moderate consuming price. We shall be making better manure and keeping more stock, because we reduce the necessity for so much hay-making.

As the cow loses her milk, the quantity of artificial food may be reduced; and if she is grazed out during the summer, the value of the pasturage will not, probably, exceed 3*s*. 6*d*. to 4*s*. a week. Thus, if we take six months' pasturage—two months for the cow being dry, and four months of artificial feeding—we have a total cost for the year of from 10*l*. 1*s*. to 10*l*. 14*s*., independent of attendance. In this calculation the straw for bedding has not been valued, or the manure made, which would amply repay both straw and attendance. Mr. Horsfall, who has gone very minutely into the questions relating to dairy management, gives us, in the pages of the *Royal Agricultural Journal*, his experience of a system of milking and feeding, which is continually practised in the metropolis. His proportion consists of a mixture of green rape cake with one quarter the quantity of malt dust, one quarter of bran, twice the quantity of bean-straw, oat-straw, and oat-shells in equal proportions. This mess, well mixed, is moistened and steamed for one hour. Cows in full milk receive in addition 2lb. of bean-meal scattered over the mess—21lb. per day, in three meals, for each cow. In addition, 28lb. to 35lb. of cabbage, kohl rabi, and mangolds, according to the season, were given daily, and after each feeding, 4lb. of meadow hay, making 12lb. a head. On such liberal diet, Mr. Horsman's cows milked largely, and gained flesh at the average rate of ¼cwt. per month, the amount of gain being in reversed proportion to the flush of milk. When in full milk—*i.e.*, for the first three months after calving—they were stationary; as the milk fell off the weight rapidly increased; and when the quantity of milk became too small to be worth taking, the animal was about fit for the butcher.

The next point, as to what may be considered a fair produce of milk and butter, is difficult to answer, depending so entirely upon the nature of the food and the peculiar disposition of the animal. In all such questions we must endeavour to get at something like an average result; and we have no hesitation in stating that well-managed cows should yield from 500 to 600 gallons of milk annually. Yorkshire shorthorns have been known to produce 800, and Ayrshires as much as 650 gallons; but these are exceptional cases. A dairy of well-bred Irish cows, as reported by Mr. Williams, of Blarney, co. Cork, yielded for the twenty-six summer weeks 384 gallons apiece, and in the winter half 198 gallons, or 582 gallons in the year. The percentage of butter depends upon food, nature of the animal, both as to breed and peculiar habits, and upon the season of the year. Shorthorns, though often large milkers, do not yield a rich article; and we may calculate that twelve quarts of milk will yield 1lb. of butter. With Ayrshires and Alderneys, nine to ten quarts will frequently suffice. Mr. Horsfall's high-fed cows (cross-bred shorthorns, already alluded to) yielded on an average 18oz. of butter for twelve quarts of milk; and each quart of cream produced 24¾oz. of butter—an extraordinary result as compared with the ordinary returns. Mr. Horsfall's cows yielded on an average 266lb. of butter per annum. The ordinary produce may be put at about 2cwt. per cow.

The question arises, What would be a fair profit under good management? The value of a cow's produce may be put roundly at a sum between 16*l.* and 17*l.* We are not now speaking of the milk trade in the neighbourhood of large towns, because, without any addition from the cow with the iron tail, new milk makes 9*d.* to 10*d.* a gallon wholesale, which leaves an average of 22*l.* 10*s.* to 25*l.* a cow. In such cases the cost of keep and attendance is greater than we have calculated, and the farm loses the value of the manuring elements, which must tell greatly in time. But, assuming that the milk will yield, whether for butter or cheese, about 6*d.* a gallon, we arrive, with the calf, at an average of about 16*l.* to 17*l.* The cost of keep and attendance being deducted, we have, barring accidents, a net return of 5*l.* to 6*l.* a cow, and something more for the value of manure over litter, which it is difficult to calculate.

With cheese and butter at their present price, will it pay to give dairy cows corn or cake when out at grass? This important question must have occurred to farmers during the last year or two, and yet we are not aware that, as a rule, the experiment has been tried. We proceed to consider the question.

The arable land of this country is yearly improving in condition under better and more liberal management, but grass land, and more especially that devoted to the feeding of dairy and store stock, is either at a standstill as far as improvement goes, or is deteriorating, and every year becoming less productive. How can it be otherwise when, as is too often the case, we are content to carry off considerable supplies of phosphates and alkalies in the produce, without supplying an ounce of foreign matter? Ultimate barrenness must inevitably terminate such treatment.

There are exceptional instances of pastures so naturally rich, having such an inexhaustible store of materials, that foreign manures are not necessary, and indeed the use of extra food would in such cases generally result in the accumulation of fat instead of milk. But upon all the medium and poorer soils cows will pay well for high feeding. Cows vary in their capacity, both for milking and laying on flesh. Some—and they are always the most useful to the dairyman as long as they are in milk—cannot be made fat. Like Pharaoh's lean kine, they swallow all before them; but, unlike them, they give a good account of their food. Such cows *must* pay for a reasonable amount of good food. Both the land and the milkpail must be enriched when it is certain that the animal is not making flesh. When cows are brought night and morning to the homestead and tied up for milking — a plan which, in all cases of central buildings, commands our warmest approval—there can be no difficulty in supplying each animal with the quantity of food proportioned to her capacity as a *milker*. In this way all have their proper allowance, which would not be the case if fed in troughs or cribs in the field. Moreover, there is a saving of labour and of loss to the food from exposure to weather.

The next point for consideration is the food that will prove most economical. Linseed oil cake, bran, or barley meal would be readily eaten, but are expensive; moreover, the former is not so well suited for the production of milk as for fattening purposes, and we think substitutes may be found which, at less cost, will produce better results. Green German rape cake has been highly spoken of by some authorities, and we believe that if finely broken, treated with boiling water, and then poured over chaff, the result would be a nutritious food. It is sometimes difficult to obtain, and animals require coaxing to eat it freely. Another food which has of late years attracted considerable attention as a substitute for linseed cake is palm-nut meal which has already been treated of. Having used this for some years, we are prepared to state that an equal quantity will increase the quality and quantity of produce more than linseed cake. The original nut contains between 55 and 60 per cent. of oil; of which about 20 per cent. remains in the meal. This oil, which is solid at ordinary temperatures, is not of a scouring nature, and in composition very closely resembles animal fat or butter. Dr. Vœlcker, who has examined a great many samples, thus speaks of its composition:—"This meal is very rich in ready-made fat. In the best linseed cake the percentage of oil rarely amounts to 12 per cent., and 10 per cent. may be taken as a fair average. The palm kernel meal contains twice as much fatty matter, and theoretically is much superior to oil cake as a direct supplier of fat. The proportion of flesh-forming (nitrogenous) matters is fully as large as in the best barley meal, but much less than in linseed, rape, or cotton cake; nor is it equal to that found in peas, lentils, and other leguminous seeds. The amount of indigestible woody fibre is but small. It contains about as much mineral matter as cereal grains, and thus is not particularly noted for bone-producing qualities." Dr. Vœlcker concludes from his investigations that the palm-nut meal is not so well adapted for rearing young stock as for fattening and milking animals, and if used for the former it should be mixed with beans, peas, or some other leguminous plant, rich in nitrogen.

So much for the opinion of the chemist. The following extract from Mr. W. T. Carrington's article on dairy farming, "Royal Agricultural Society's Journal," vol. 1, second series, p. 347, gives the result of practical experience:—"At the large cheese fair held annually at Leicester early in October, Mr. Nuttal, of South Croxton, pitched 12 tons of cheese, of first-rate quality, which was sold altogether at over 8d. per pound. This was that portion (about two-thirds) of his yearly make from some eighty-five cows, which was ripe at the beginning of October. He attributed his large make of cheese in that unfavourable season (1864) in a great measure to the use of 30 tons of palm-nut meal, of which he spoke very highly." We have known cows in full milk, eating 3lb. to 4lb. a day, produce as much as 4lb. of cheese. Why, then, it may be asked, is not this meal in more general use? Principally because some difficulty occurs at first in inducing animals to eat it. The meal is dry and gritty, and too often masters leave details entirely to ignorant servants, who condemn without a fair trial. We have not yet heard of a single case of failure where ordinary patience was exercised; and when once animals take to it they will consume it freely. The winter time is the best season for commencing its use, as fatty foods are more acceptable at that time than in summer. It has been suggested to the manufacturers that they might with advantage introduce some 15 or 20 per cent. of ground carob beans to give a flavour; but they wisely decline to do so, as compromising the purity of their article. This, or other suitable mixture, may be made by the consumer, and will be found valuable, especially in summer. Treacle, when cheap, might be used. In this way our cows may eat 3lb. to 4lb. of such a mixture with great advantage, as we are convinced it would materially influence both the quantity and quality of the milk, more especially the latter. On all mixed arable and dairy farms there is generally a considerable quantity of tail or damaged corn, which may be ground and mixed with the palm-nut meal. Ground oilcake would answer the same purpose, but would be more expensive, though probably producing more cream or curd than the same weight of barley or wheat meal. Taking the cost of palm-nut meal at 7l. a ton, linseed oilcake at 12l., and barley meal at 9l. a ton, and comparing the cost of the two latter, we find that 3lb. of palm-nut meal, and 1lb. of linseed or 2lb. of barley, would cost about 3$\frac{3}{4}d$. a day, or 2s. 2$\frac{1}{4}d$. a week. Supposing one cow so fed from May 1 to October 1, she would consume food to the value of 48s.—not a very serious item, and well paid for in the extra produce and improved land. Some of our readers may perhaps condemn this idea as theoretical; but if they will only give it a fair trial, they may soon see that it will pay. It is proved beyond doubt that feeding cattle pay for cake, even with beef at a much lower price than it now is; and why should not dairy cows pay for good treatment, when cheese is worth 80s. per cwt., and butter 16d. to 18d. a pound? The more stock a farm can be made to carry the better, providing we have food enough, and can maintain them in a healthy, thriving state. Cows receiving 4lb. of meal a day would do with a third less pasturage, and where dairies are already large, the increase in the number of cows which the land, with assistance, would be able to carry, would appear considerable, and would greatly affect the cheese tub.

A word or two in this place as to the virtue of cooked food. The actual gain by this process has never been satisfactorily tested, nor how far the extra cost is paid in increased results. There are many practical men who strongly advocate the practice. Mr. Horsfall was one of the first to carry it out on a large scale with his cows, and we learn from Mr. Edmonds that the late Mr. William Bowly, after careful trial, adopted the following plan: The quantities are given for fifteen cows; a furnace containing 70 gallons of water, the water hot to the boiling point; then meal at the rate of 10lb. per cow, to be well stirred in and boiled gently for an hour. Half of this to be poured over chaff (three bushels per cow), placed in a long trough for the morning's meal; the remainder being used in the same way in the evening. The chaff and soup are thoroughly mixed, and left for about half an hour to cool before being used. One great point is to have the mixture as fresh as possible; all food that has been cooked is apt to turn sour if kept beyond twenty-four hours. Mr. Edmonds himself, who farms extensively, is an advocate for cooked food, recommending a large proportion of hay—two parts out of three during the early stages, and, later on, three parts of hay to one of straw—and he goes on to say that with this he gives 4lb. or 5lb. of oil-cake, and half a peck of mixed meal, barley, and peas or beans, also a proportion of wheat, to be increased to one peck a day as the beast advances. The oil-cake and meal to be boiled in water for half or three-quarters of an hour, and thrown in the form of rich soup over the chaff, well mixed and salted. On such food, with the addition of a peck of roots daily, Mr. Edmonds makes some capital Christmas beef, and we should be much surprised if he did not, as the diet is highly nourishing.

The plan of using so large a proportion of hay may be justified in Mr. Edmonds's case by peculiar considerations, but as a practice must be denounced as extravagant. No doubt the perfume is fragrant, especially when the soup is first mixed, but the question is how far the nutritive properties of the food are increased. It would be very serviceable if Mr. Lawes, that prince of experimenters, would test the point with two lean animals, supplying the same amount of food, and weighing and analysing the excreta. If the digestive process is really so much assisted as some suppose, then a large proportion of nutriment would be extracted, and the dung would be poorer; the progress of each animal being ascertained by frequent weighing. Mr. Edmonds dwells upon the necessity of having the food fresh. "The mixture with the chaff should never be made more than twelve or fourteen hours before being used; seven or eight hours is better." Another plan is to place the chaff in a bin, and pour the soup over it in layers. The heat is thus kept in longer, and the fermentation is possibly greater. Lastly, waste steam is frequently passed through the chaff, roots, and meal, and the whole partially cooked. This is particularly desirable in case the fodder is inferior; mouldy hay is much sweetened and made more palatable. Food so treated must be presently used, or it turns sour and is not so readily eaten. The success of the cooking system depends mainly upon good management and constant supervision. Men are apt to become careless, and omit to thoroughly clean out mangers or coppers, or allow the food to become sour, and thus upset the appetite. Taking all points into consideration, and until convinced by actual experiment, we are in favour of giving the food uncooked, as the simpler process, and one that answers remarkably well.

In the next place we have to consider the arrangements of the dairy and details of management.

A regular and easily regulated temperature is of great importance; hence the dairy should face north, and be sheltered from the south. It must be well ventilated. This can be secured by carrying the walls up, introducing under the ceiling a row of ventilating bricks, and having the space occupied by the movable casement covered with perforated zinc. We thus keep the temperature equable in summer, and in winter a hot water apparatus, with circulating pipes, secures the requisite warmth. A dry atmosphere is desirable; hence, sinking the dairy beneath the surface, thereby insuring dampness, is not recommended, though often done. We should prefer taking the ground level, building the walls with a damp course, and laying the floor, which should either be flags or brick squares, on concrete. A good drain round the building is desirable. If the bricks are machine-made, and sufficiently even to make a good face, we should make the walls 14in. wide, with a hollow space of $4\frac{1}{2}$in. inside, and point both inside and out; we thus avoid the necessity for plaster, which takes a long time to dry, and until dry is likely to affect the milk. Three coats of Carson's iron paint makes a good finish inside, filling up all spaces in the bricks, and leaving a surface that can be easily cleaned, and on which nothing offensive can lodge. It should always be remembered that milk is subject to fermentation, and that it turns sour rapidly; hence, both in the dairy fittings and utensils, we should have surfaces into which moisture cannot soak. Where outlay is no object, the walls are often covered with glazed tiles, which look remarkably well, and are easily cleaned. The glazing must be well done, otherwise there will be risk, even with them, and we prefer painted walls, as infinitely cheaper. The shelves, or dressers, on which the milk bowls stand, may be either of slate or wood. We much prefer the former; but if the latter are adopted on the score of economy, Dr. Vœlcker's advice as to paint is excellent, as we thereby prevent milk accidentally spilt being absorbed by the wood. Three coats of best white paint, and this renewed frequently, will answer the purpose. Slate looks very neat, and is easily cleaned, but the slates must be in one piece, and not jointed, or only in such a manner that the joints are moisture-proof.

We have said that the walls should be high, the roof covered with red tiles, if straw cannot be used; the latter insures the most even temperature, and is the coolest in summer. Black slates should be avoided, or, if they must be used, straw should be packed underneath; but this is objectionable, as harbouring rats and mice. We have a considerable choice of vessels. Slate cisterns can be recommended only when cut out of the solid block; they are then non-absorbent, and readily cleaned. If jointed, as is often the case, the milk is absorbed by the cement, and will not keep. Glazed earthenware pans are cheap, and answer well, provided the glaze is even and good. Unglazed pans or vessels of wood are most objectionable, for reasons stated; glass is clean, but liable to break. Dr. Vœlcker refers favourably to the tinned iron vessels, showed by Major Gussander at the Exhibition of 1862. These vessels have a small opening at the bottom, fitted with a brass plug; over the opening a cylindrical tube provided with narrow slits is soldered. When the plug is raised the skim milk escapes, leaving the thick cream, which cannot pass through the narrow openings at the bottom of the pan. They should not exceed 2½in. in depth, as a loss of cream ensues if the milk stands in a deeper vessel. Another advantage in using shallow metal vessels is that the temperature of the milk is more rapidly reduced, especially if the vessels stand upon slate or stone shelves. It is very important that this reduction of temperature, say from 90°, at which it comes from the cow, to about 60°, should be as rapid as possible. Dr. Vœlcker believes that in a proper dairy twenty-four hours is sufficient time to leave the cream to rise, and he quotes an experiment of Sannert's in proof. He left two equal quantities of milk, the one for thirty, the other for sixty hours before skimming; the former produced 30lb. of butter, the latter only 27lb.

Mr. Horsfall, whose papers on Dairy Husbandry, first published in the Journals of the Royal Agricultural Society, are most valuable, gives the following description of his own dairy:—Dimensions, 6ft. wide by 15ft. long, and 12ft. high. At one end (to the north) is a trellis window, at the other an inner door, which opens into the kitchen. There is another door near to this which opens into the churning room, having also a northern aspect. Both doors are near the south end of the dairy. Along each side of the north end two shelves of wood are fixed to the wall, the one 15in. above the other; 2ft. higher is another shelf, somewhat narrower, but of the same length, which is covered with charcoal, whose properties as a deodoriser are sufficiently established. The lower shelves being 2ft. 3in. wide, the interval or passage between is only 1ft. 6in. On each tier of shelves is a shallow wooden cistern, lined with thin sheet lead, having a rim at the edges 3in. high. These cisterns incline downwards slightly towards the window, and contain water to the depth of 3in. At the end nearest the kitchen each tier of cisterns is supplied with two taps, one for cold water in summer, the other with hot water for winter use. At the end next the north window is a plug or hollow tube, with holes perforated at such an elevation as to take the water before it flows over the cistern." In summer the kitchen door is closed, and an additional door fixed, the space between being well packed with straw; a curtain of calico, kept moist with salt and water, is hung up before the window. The water in the cisterns is so arranged as to trickle slowly through all the time, and thus the temperature is reduced and the milk remains sweet. In order to reduce the temperature of the cream so as to insure firm butter, Mr. Horsfall hit upon a simple and successful plan, viz., to lower the cream about 28ft. down a well, and leave it there the night before churning. He had a windlass made for this purpose, and the cream jar placed in a basket suspended from the rope. So great is the advantage thus acquired, that sooner than be without it, Mr. Horsfall would go to the expense of sinking a well. We mention this fact as important to bear in mind in constructing a new dairy. A supply of good water is essential, and it would be easy to sink the well either under or adjoining the dairy, so as to be within reach. Scrupulous cleanliness is necessary, but this rather by the keeping out dirt than by the constant use of water. All vessels must be scalded out immediately after using, the walls, floors, and dressers wiped over with a damp cloth, and speedily dried. Nothing is so injurious to the keeping of milk as a damp atmosphere. The atmosphere should be sweet; bad smells from accumulation of manure, cesspools, &c., are very injurious, and should be guarded against in the selection of a site.

It would be impossible in a brief sketch like this to enter into a lengthened description of cheese-making, neither is it so necessary now as formerly, since the principle of co-operation in dairy management appears to be steadily gaining ground. Dr. Vœlcker, who prosecuted careful researches at the request of the Royal Agricultural Society, published the result of his investigations in the twenty-second and twenty-third volumes of the Society's Journals. Anyone who is about to commence cheese-making will do well to study these papers. Dr. Vœlcker is convinced that the food has undoubtedly an influence on the quality of cheese: the method of manufacture is still more important in determining the quality and character, and one great argument in favour of the Association Dairy system is that uniform and scientific management is adopted, and though the milk must vary considerably, as coming from land of varying fertility, the produce generally commands the top prices in the market. The first idea of a cheese factory was derived from America, where the system is general, and the great and rapid improvement that has been made during the last few years in the quality of American cheese is undoubtedly due to more scientific treatment. The experiment in this country was tried first in Derbyshire, where factories at Longford and in Derby have been in operation since 1870; and notwithstanding an array of prejudice and trade opposition, there is every prospect of a successful issue. The saving of drudgery to the farmer's family in escaping from the details of cheese-making, will endear the system to farmer's wives, whilst the economy of labour must give better results in the long run than could be looked for from private enterprise.

Milk consists of curd or casein, butter or fat globules (milk sugar), and mineral matters. In the preparation of cheese, the curd or casein is separated by means either of lactic acid, which forms when the milk becomes sour, or by the addition of rennet; the butter is also separated if we want rich cheese, and a portion of the mineral matter; whilst in the resulting fluid, known as whey, remains the milk sugar, and most of the mineral matter. The Cheddar process, which is the safest and most largely followed, is thus described by Dr. Vœlcker:—"Immediately after morning milking, the evening and morning milk is put into a Cockey's tin tub, having a jacketed bottom for the admission of steam or cold water. The temperature of the whole is slowly raised to 80°, by admitting steam into the jacketed bottom. The rennet is now introduced, the tub covered with a cloth, and left for an hour. If annatto for colouring is used, it must be added before the rennet. Good rennet should properly coagulate milk at 80° in from three-quarters of an hour to an hour. If the milk fail to be coagulated within the hour, the curd produced will be tender, and not easily separated from the whey without loss of butter; whereas, on the other hand, if the curd is separated in twenty to twenty-five minutes, the cheese is usually sour or hard. Great care should be exercised, in preparing the rennet so as to insure uniform strength. At the end of the hour the curd should be partially broken, and allowed to subside for half an hour, after which the temperature is gradually raised to 108° Fahrenheit, the curd and whey meanwhile being gently stirred with a wire breaker, so that the heat is uniformly distributed and the curd minutely broken. The heat is maintained at 108° for an hour, during which time the stirring is continued. The curd, now broken into pieces the size of a pea, is left for half an hour to settle; at the end of this time the whey is drawn off by opening a spigot near the bottom of the tub. As the curd should be quite tough, no pressure is at first requisite to make the bulk of the whey run off in a perfectly clear state. The curd now collected in one mass is rapidly cooled, cut across into large slices, turned over once or twice, and left to drain for half an hour. As soon as it is tolerably dry, it is placed under the press, and most of the remaining whey is removed by pressure. After this the cheese is broken, first, coarsely, by hand, and then by the curd mill, which divides it into small fragments. A little salt should now be added and thoroughly mixed with the curd. The next operation is vatting. The cheese vat, carefully filled with the broken and salted curd, is covered with a cloth; the curd is reversed in the cloth, put back into the vat, and placed in the press. The cheese cloth should be frequently removed, and the cheeses are ready to leave the press on the sixth morning.

This is the process according to the Cheddar system. If carried out as described, and after treatment carefully attended to, viz. the turning and wiping of the cheeses and the maintenance of the cheese-room at an even and somewhat high temperature, the result will be a well-made eatable cheese, varying as to quality according to the nature of the food.

The following account of continental dairy management was contributed to the *Field* by a valued correspondent, and is here inserted as illustrating the variety of practice successfully pursued. It is an account of the system pursued by Mr. Schwartz, of Hofgarten, on Lake Wetter, which has found favour in Denmark, Sweden, and many other parts of the Continent; and, although in some respects it differs radically from the practice pursued by dairymen in this country, it may still be found to merit careful attention, especially by those with whom economy of space is an object.

Before touching upon the characteristic features of the system in question, it will perhaps interest the reader to hear that the farm occupied by Mr. Schwartz consists of 1200 acres of arable and 107 of pasture and meadow land, and that the method of culture followed is the one known as the Holstein "Koppelwirthschaft"—a ten or eleven years' course of cropping, in which wheat, barley, and oats are the chief cereals, and timothy grass, hop trefoil, and clover the intervening forage plants.

The live stock includes 160 to 170 cows ("herrschaftliche," or well-bred animals, and shorthorn, with their respective produce), 20 oxen, and 32 working horses.

The cows remain out during summer, but are driven under cover at nights. 732 head of dairy cattle are found to consume 2¾ acres of green food a day, and the herdsmen mark off every morning, by means of a "mahde," or way cut with the scythe, the area over which the cows may roam. It rarely happens that they overstep this boundary, though not unfrequently they may be seen wandering back to the previous day's bite—a liberty freely allowed them. To the green food rations a somewhat curious addition is made—viz., horse-dung and split peas. Mr. Schwartz, it appears, had read years ago, in some old Swedish muck-manual, that horse-dung, mixed with other provender, might be advantageously used as a feeding material for cow stock; and noticing that a certain shorthorn bull, even after a hearty meal, would frequently eat the horse-dung collected in a corner of the farm-yard, it struck him that it might be as well to give so palatable an article of dessert a systematic trial as a condiment, if not *pièce de résistance*. He accordingly instituted a series of experiments with the new food, and so satisfactory was the result that he ended by giving his cow stock daily as much as 8lb. per head of it along with their split peas. He found the dung had a favourable effect on the amount of butter in the milk, and peas on that of casein. That cattle, after luxuriating all day on good pasturage, should fall upon the above compound with avidity, has surprised more than one visitor to Hofgarten. Facts, however, are stubborn things, and Mr. Schwartz, thanks to the introduction of the new feeding material, has even found it possible, on the same acreage, to keep twenty to thirty more cows than he was formerly able to do, and still adhere to his rule of giving them all through the year as much food as they will eat. The loss which the land sustains in the way of horse-dung is more than compensated for by greatly increased dressings of manure from the cow stables.

To turn now to the dairy arrangement proper, the milking, we should observe, is done in the stalls by women, and the milk emptied at once out of the pails into large copper vessels immersed in water. It is afterwards removed in carts to the milk room, and there we first become acquainted with one of the leading features of the Schwartz system—viz., that of letting the cream rise at the lowest temperature possible. When the milk has been measured in a cylindrical tin vessel, with a glass scale inserted in its side to indicate the quantity, it is poured into tinned-iron cans, 20in. high by 16in. or 17in. broad; and these, with their perforated lids on, are suspended in a tank occupying the middle of the room.

During summer the water is kept by means of ice at 39° or 40°, or at most 44° Fahr.; whilst in winter its temperature ranges but little above freezing point. The method of preserving ice adopted at Hofgarten is at once simple and effective. In an open space, at some little distance from the farm premises, a tolerable-sized hole is dug, and into it (to allow the water to flow off) a flooring of wooden trellis-work is laid. The labourers then pile up the blocks on this foundation, and, having carefully filled all the interstices, pour water over the heap, in order that it may freeze into a compact mass. They next cover the miniature iceberg with a thick layer of sawdust, and finally close in the whole with a substantial straw thatch. To obviate the necessity of opening the ice-house at frequent intervals, a considerable quantity of ice is taken out at a time, and placed in the immediate vicinity of the dairy. Protected like the greater mass by a good covering of sawdust, practically but little of it melts away, even at a temperature of 70° to 80°.

The buildings at Hofgarten devoted to cheese and butter making are handsome in appearance, and at the same time practically arranged. A horse-gear, which is used for churning, &c., occupies a space near the chief entry, and, stepping into the dairy, the eye is first attracted by a boiler and a number of utensils undergoing a thorough cleaning at the hands of the head dairywoman and her assistants. Not only in this, but in all other branches of the establishment, the work is performed entirely by females, and everywhere the most scrupulous cleanliness prevails. In cases—necessarily arising in the manufacture of butter and cheese—which demand delicate handling, judgment, and great exactitude, the head dairywoman, single-handed, undertakes the operation. Adjoining the so-called wash room is situated the curd kitchen, fitted up with four rectangular boilers of English make, four feet in height and diameter. In order that the milk may be raised to the proper temperature without coming in direct contact with steam, these boilers are jacketed and surrounded (at a distance of half a foot) with hot water, and above the spigot in front of each of them a perforated tin plate is introduced, to prevent particles of curd escaping with the whey.

Coagulation having been brought about by the addition of rennet, the mass is stirred round for twenty minutes with a curd-breaker, and afterwards, at a temperature of 84° Fahrenheit, cut into small pieces by means of knives attached to an upright shaft in the boiler. The knives are moved backwards and forwards by horse-power, half of them cutting horizontally, half vertically. In about twenty minutes the hot water surrounding the boiler is found to have raised the temperature to 98°, and this to-and-fro movement of the mass is continued for another half-hour or so. The spigots are then turned to allow the whey to drain off, and any buttery particles which may escape with the latter are afterwards skimmed off and utilised for butter making. As regards the curd, it is once more cut small, packed in deep perforated tin-plate moulds, and removed to the press rooms. These, situated above the offices already alluded to, are furnished with twenty superior English presses, capable (each) of holding two cheeses, and adjustable to any required degree of pressure. After having wooden covers fitted closely over them, the tin moulds containing the cheeses are piled on one another in twos, with a board betwixt them, and placed in the presses. At an early hour of the following morning they are taken out, cut afresh into pieces, and cooled down to a temperature of 55° Fahrenheit. About midday they are broken and crumbled in a curd mill; and the next operation, previous to their being thoroughly kneaded by the head dairywoman and again exposed to pressure, is that of salting them. The salt requires, of course, to be distributed as uniformly as possible, and the time during which the cheeses remain under gradually increasing pressure, varies from three to four days. They are then removed to the cheese room; and by the end of two or three months, if the different operations above described have been carefully attended to, they will have acquired a delicate flavour and be quite ready for market. Those intended for export to England are sewn up in linen cloths, to protect the rind from injury. They sell in Hofgarten, under the name of "fat Cheddar," at 7½d. to 8d. the Swedish pound (15oz. English). On an average, it is found that 620lb. of milk yield in summer 62lb. of new—*i.e.*, just pressed—and 53lb. of dry or ripe cheese.

The cheese room at Hofgarten—a detached building, partly under, partly above ground — testifies to the many experiments in cheese-making which occupy the attention of Mr. Schwartz. Arrangements have been made for heating it, and thus complete protection is afforded against the influences of the outer air. The shelves, ranging above one another to a height of twenty feet, are easily reached by means of a movable floor, and, owing to the loftiness and construction of the cellar, the cheeses can be placed either in a dry or moist atmosphere, and in the temperature best suited to them. They have each of them, whether English or Swedish—for Mr. Schwartz makes, in addition to so-called Cheddar, fat, medium, and thin Swedish cheeses—a label attached, indicating their weight when taken from the press, the date of completed manufacture, and the quantity of milk used.

As regards the making butter, this operation is conducted at about 50° Fahr., and, as before mentioned, the maintenance of a low temperature (especially during the rising of the cream) and the employment of deep milk pans, or rather cans, form two of the chief characteristics of the Hofgarten system. Mr. Bliss, director of an American butter factory, recently observed that "old theories to the contrary notwithstanding, cream will rise as readily through one foot as one inch of milk," and Mr. Schwartz's experience would almost bear out the assertion. Repeated trials and carefully recorded experiments have shown the following to be the rate at which it is thrown up at a temperature of 50°, the quantity of milk stood being 144lb.: During the first twelve hours, 5·30lb.; during the second twelve hours, 0·17lb.; and during the third, 0·06lb. As nearly all the cream thus rises during the first twelve hours, it is not considered worth while to prolong the process of separation beyond that time. Whatever has not then risen serves to enrich the thin cheese made of the residue. The churn used is of a barrel shape, with upright axis, and worked by the horse-gear.

When intended for export to England the butter is packed in casks holding 60lb., and these, before being used, require to be carefully rinsed out and (whilst still moist) sprinkled with salt. The layers of butter do not, however, come in direct contact with the sides of the casks; a linen cloth, which has been steeped in brine, is first placed round them. Much importance is attached by the English trade to these matters of detail, and Mr. Schwartz had difficulty in disposing of his much-approved-of dairy produce until he complied with the request to cover it with cloths of a special quality.

Enumerating, in conclusion, the advantages claimed for the deep-can and low-temperature system exemplified at Hofgarten, we should mention, first and foremost, the greater economy of space and vessels it permits of. Whilst holding a much larger quantity of milk than the ordinary pans or dishes, the deep cans recommended by Mr. Schwartz occupy, proportionately speaking, far less room; and, owing to this circumstance, the farmers of Sweden and Denmark who have taken to them find that extensive dairy premises are no longer needed. Even on holdings of a considerable acreage, a small or but moderate-sized wooden construction, with thatched roof, is all that is now deemed necessary. A second point in favour of deep cans is the comparatively short time in which they can be cleaned. The surface that requires washing is of course less than when shallow vessels are used, and there are a fewer number of corners. Being provided with lids, the milk contained in them is protected against dust or any injurious matter floating in the air, and thus greater cleanliness is secured. Another circumstance favouring the use of them is the facility with which the cream, in consequence of its greater depth, can be skimmed without removing any of the milk below. With regard to the advantage—a twofold one—that results from the maintenance of the milk at a low temperature, it consists in obtaining the cream, and also the skimmed milk, in a perfectly sweet condition, and reducing the time the former occupies in rising to twelve hours.

We leave our readers to judge of the value of the two systems. Probably each is suitable to the condition for which it was designed.

Of late years a great impetus has been given to the milk trade owing to the increased demand, and the facilities for transport offered by railways. The following account of a milk farm gives a general idea of management. Moreover, the subject has peculiar interest, inasmuch as public attention has been very properly called to the disgraceful adulteration that has been so commonly practised, and we trust to be able to show that milk can be produced, and profitably sold in a country town, at the moderate rate of 1s. a gallon.

The farm we have in view is situated within two miles of a manufacturing town in the Midland Counties, and comprises about 500 acres, of which 200 are arable—a rich loam varying in consistency from quite light sand to strong soil, on the old red sandstone. No particular rotation is adhered to. Generally, however, a three-course shift is adopted, namely, wheat, mangolds, and pulse, *i.e.*, a mixture of beans, peas, and oats. Seeds are not wanted, as the pastures afford abundant food. Farmyard manure is applied to roots, being spread upon the stubbles after harvest, and buried with a deep furrow. The great object in this rotation is to produce the most paying crops; thus wheat yields, by corn and straw, more than any other cereal. Mangolds are especially valuable for milk, and the pulse crop affords a great weight of grain, which, as meal, is found conducive both to milk and flesh, whilst a large proportion of the straw is also edible. Both the wheat and the pulse crops are stimulated by dressings of artificial manures in the spring.

The London cowhouse system is adopted. Cows purchased just before calving, principally from the dales of Yorkshire, are tied up in sheds as soon as they have calved, and never go out again until they are fit for the shambles—being moved on from shed to shed according to their yield of milk, the object being to milk and feed at the same time. A good milker, passing more of the food into the bucket, remains in the earlier sheds a longer period, and feeds more slowly. The difference in this respect is very remarkable—extraordinary milkers having been kept upwards of two years; the average, however, ranges from ten to twelve months, at the end of which time the cows are sufficiently fat to be sold by auction; when, if they realise as much as they cost in, it is considered satisfactory. In winter the food consists of pulped mangold (80lb. per head), with chaff composed of equal portions of hay and straw, supplied three times a day. The artificial food, which is distributed over the chop, consists of a mixture of palm-nut meal, bean and pea meal, and linseed cake, averaging 4lb. per head daily—the linseed cake being principally used in the later stages. Both palm-nut meal and pulse meal are very highly spoken of for their influence both upon the quantity and quality of the milk. In addition to the above, each cow receives daily a bushel of grains, costing 4d., which are also mixed with the pulp, and a feed of cabbage supplied whole. The cows are bedded with sawdust, sprinkled lightly over the bricks. They are carefully dressed over, curried, and brushed once every day, which is found absolutely necessary to promote healthy circulation. The cattle are fed from a headway; water is laid on in pipes, but not supplied in troughs, as the mangers are kept cleaner, and the water is fresher when supplied in buckets; once a day is found sufficient. The feeder has half a dozen buckets, which he fills from large-sized taps in the gangway. The process is much more expeditious than appears from our description; it takes less time than would be expended in turning out and tying up again, and is far more satisfactory in every respect, as the cows do not become restless, and the risk of injury from goring is avoided.

During the summer—*i.e.*, from April to October—the soiling system is pursued with marked success. Rye and early vetches supply food before the permanent grass is ready; these are given in conjunction with the last of the mangolds. The mowing ground, occupying about 200 acres, is liberally treated. As soon as the first crop is removed, a dressing of the cow manure, at the rate of about six or eight loads per acre, is applied; carefully spread and well worked in with chain harrows. The absence of straw, and the substitution of sawdust, renders the manure very applicable; and with an ordinary amount of rainfall it rapidly disappears, and does not in any way, save beneficially, affect the second crop. It is only the earlier mowings that produce a second crop. About 100 to 120 acres are reserved for hay. As far as is practicable, foldyard manure is applied to all the mowing land once in two years, and alternated with composts of soil and lime, which latter has a marked effect upon the leguminous plants. Similar proportions of artificial food are given in summer, the grains forming a medium for their distribution. As the season advances, and the grass becomes hard and dry, it is found desirable to cut it into chaff, which prevents waste. Great attention is paid to hygiene. Thus the sheds are frequently whitewashed, and daily sprinkled with a powerful disinfectant. By means of carefully-arranged ventilators an even temperature, summer and winter, is maintained, which is found to have a material effect upon the yield of milk.

Cases of milk fever are rare — attributable to the great care in reducing the cows before calving. When the animals are within a day or two of their time, they are placed in a loose box, and allowed only a mouthful of hay or grass, with water. The calf is removed at once, and when one or two days old is sold, the prices for the past year having ranged from 2*l*. 10*s*. to 3*l*. On the average eighty cows are kept, of which about half a dozen are being dried off for the market, as it is found that they sell so much better when the udder is properly shrunk.

Four women and two men are employed milking, and the head cowman overlooks, and strips the cows. Work commences at 4.30 a.m., and is finished by six. The milk is removed from the buckets into large cans, which in summer are placed in a trough under a cool shed, and half covered with water. This is done in order to reduce the temperature as rapidly as possible, and thereby prevent the milk from turning sour. The afternoon milking commences at 12.30 and ends at two o'clock, this unusual arrangement being necessary in order to supply the market in time for the ladies' kettledrums and the children's teas.

Here, according to ordinary routine, our description would cease, as the retailer is introduced on the scene; but in the case before us the milk-grower is the milk-seller, having a dairy with a manager, and milk-walks employing six men. The milk is conveyed from the farm in a spring cart, and delivered to the manager, who is responsible for its sale. Immediately on its arrival, the milk is measured out to the different salesmen, according to the requirements of each district, and booked against them. On their return they account to the manager for cash receipts and book customers.

A journal is kept at the dairy, in which the rounds of each man are entered and where he ought to be at particular times; thus an oversight is maintained, and customers are solicited to forward complaints to the manager in case the quality of the milk is not up to the mark; fraudulent transactions on the part of the men are thus guarded against. The milk is conveyed in cans placed in three-wheeled trucks, which are pushed along.

A portion of the grass land is kept as pasture, affording a stray for the cows when they first arrive, and also helping to carry a flying flock, consisting of one hundred ewes, purchased in the autumn, which, with their produce, are sold out during the next summer or autumn.

Milk-producing farms are liable to deterioration, owing to the constant removal of produce rich in mineral elements. This is met in the case we are describing by the large outlay in artificial food, which amounts to over 500*l*. a year, a great part of the manure from which is applied to the grass land; and all the arable crops are liberally dressed with hand manures.

It should be mentioned that the farm, being within two miles of the town, commands a high rent, averaging with rates and taxes over 3*l*. an acre; and, considering that genuine milk is retailed at three-pence a quart, the fact that a profit has been obtained is an evidence of admirable management, and proves the remunerative character of the business when properly carried out. Considering the increased cost of production, which prevails in all departments, beginning with the cost price of the cows, the labour of attending them, and the food they consume, it is evident that the good people of Worcester obtain a genuine article considerably below its ordinary market value. We should have thought 4d. a quart for well made milk quite a reasonable price. In the neighbourhood of Brighton we are acquainted with a large and successful farmer who keeps a considerable herd, and sells to the retailer, who fetches the milk from the cowshed, at 3d. a quart. In London the retail price of milk is 5d. a quart, and we believe those who supply the milk get about 3d. a quart, carriage paid, so that the retailer has 2d. a quart to pay him for the expenses of distribution, profit, &c. A great improvement in the average quality of London milk has resulted from the penalties inflicted upon dishonest tradesmen. The new Adulteration Act promises to confer great advantages upon the poor in towns, by insuring them a genuine article, and only those who have suffered can form a just opinion of the importance of good milk.

PART II.
THE GENERAL BREEDS OF CATTLE.

CHAPTER VI.

SHORTHORNS.

BY JOHN THORNTON,
EDITOR OF "THE SHORTHORN CIRCULAR."

THE SHORTHORN BREED OF CATTLE may be fairly called cosmopolitan. Its habitat is everywhere. From one small spot in Britain, its native home, it has spread through this country till it is found from John o' Groat's to Land's End; everywhere in Ireland it prevails; to most parts of the globe it has emigrated; and the present year opened with an importation of three Shorthorn heifers and a bull by the Government of Japan.

The "art and mystery" of breeding has worked marvels upon our native breeds of sheep, and upon Shorthorns the modelling powers of man have been so exercised that the gaunt, ungainly form has been fashioned into a parallelogram of symmetry and beauty. There seems little doubt that from time immemorial the breed existed as a local race along the rich grazing valleys of the Tees, in the counties of Durham and Yorkshire. Noblemen and squires, with a thoroughly English love of good stock, kept up the herds on their estates with as much pride as their own pedigrees. Numerous are the local records of the excellencies and feeding properties of these cattle, and of their capability of attaining enormous weight when at full maturity. Mr. Chas. Colling, of Ketton, county Durham, a man of great judgment and sagacity, and a follower of the great Bakewell, was the first to bring the breed prominently into notice. He collected the best specimens together, and by careful selection and breeding reared a herd of fine cattle, which arrived at maturity much earlier than had previously been known. His brother, Mr. Robert Colling, of Barmpton, was also an eminent breeder, and followed in the same course. Both brothers at first used the same bulls. The former bred an ox of wonderful dimensions, whose live weight was 34cwt., and the latter a white heifer of equal celebrity. The bull Favourite (252) was their sire. They travelled throughout England, and were exhibited in London, as well as at country towns. This circumstance, coupled shortly afterwards with the sale of Mr. Chas. Colling's herd in 1810, at an average of 151*l*. 8*s*. for forty-seven head, brought the breed into notoriety, and from the beginning of the century Shorthorns began to spread until they may now be found in every county.

The breed is distinguished by its symmetrical proportions, and by its great bulk on a comparatively small structure, the offal being very light, and the limbs small and fine. The head is expressive, being rather broad across the forehead, tapering gracefully below the eyes to the open nostrils and fine flesh-coloured muzzle. The eyes are bright, prominent, and of a particularly placid, sweet expression, the whole countenance being remarkably gentle. The horns (whence comes the name) are usually short, springing well from the head, with a graceful downward curl, and of a creamy white or yellowish colour, the ears being fine, erect, and hairy. The neck, moderately thick (muscular in the male), and set straight and well into the shoulders, which when viewed in front are wide, showing thickness through the heart, the breast coming well forward, and the fore legs standing short and wide apart. The back, among the higher-bred animals, is remarkably broad and flat, the ribs springing well out of it, barrel-like, and with little space between them and the hip bones, which are soft and well covered. The hind quarters are long and well filled in, the tail being set square on to them; the thighs meet low down, forming the full and deep twist; the udder not too large, but placed forward, the teats being well formed and of a medium size; and the hind legs standing short and straight to the ground. The general appearance should show even outlines. The whole body is covered with long soft hair, there frequently being a fine undercoat, and this hair is of the most pleasing variety of colour, from the soft white to the full deep red. Occasionally the animal is red and white, the white being found principally on the forehead, underneath the belly, and a few spots on the hind quarters and legs; often the whole body is white, with the neck and head partially covered with roan, while again the entire body is most beautifully variegated of a rich deep purple or plum-coloured hue. On touching the points, the skin is found to be soft and mellow, as if lying on a soft cushion. In animals thin in condition a kind of inner skin is felt, which is the quality or handling indicative of those great fattening propensities for which the breed is so famous.

The Ketton herd was of this character, the knuckles or shoulder points being rather strong and somewhat upright. The red colour of some of the cattle was of a yellow tinge, and this may still occasionally be seen. Many of the earlier breeders sought to remedy the defects that some thought apparent in the Ketton stock, notwithstanding the public approval, as shown by the 151*l*. average, and the 1000 guineas for the six-year-old bull Comet (155). Mr. C. Mason, of Chilton, Durham, improved the shoulders of the cattle in his herd, somewhat at the sacrifice of the hind-quarters; and Sir Chas. Knightley, in more recent years, was also very particular about the fore-quarters. During the last quarter of a century fashion has run high, and there has been a constant adherence on the part of breeders to particular strains of blood, producing different characteristics of the same breed. In some strains style and elegance have been successfully cultivated; the beautiful head is carried erect, the horns incline upwards, the body has become elongated, and the shoulders have somewhat retained their uprightness; the whole animal bearing a most stylish attractive look. Animals of this type generally possess good milking properties, abundance of soft rich-coloured hair, and thin touch. By other breeders massiveness and symmetry, with sloping shoulders and a great disposition to heavy flesh, have been studied and attained. The adjudications in our recent showyards have been made to animals of the greatest substance, whose form has nearest preserved the type of the earlier Shorthorns, without the coarseness of their shoulders.

The system pursued by the most eminent breeders is that of allying animals of the same strain of blood. One breeder readily sells his bull calves for 500 guineas each, and a calf sold at the close of last year made 1000 guineas. Another lets his bulls for the year at varying rentals, from 100 to 300 guineas each. The most effective method of producing good animals seems that of close breeding, or as it is termed in and in, so long as robustness, size, and constitution are maintained. Another method is to use bulls of one strain year after year, upon a herd of different blood. Attention to pedigree is found to be more efficacious than attention to form without pedigree; but breeding to fashion, and the stimulating effects of high prices, have a strong tendency to deter selection and vigorous weeding, though they have likewise the tendency to perpetuate pure blood, and to avoid the deleterious effects of rash and injudicious crossing.

The Herd Book has been the mainstay of this carefulness on the part of the breeder. Brought out in 1822, by Mr. George Coates in his old age, it was continued by his son, at whose death it was taken up by the present editor, Mr. Strafford, by whose persevering labours it has reached its nineteenth volume, which was published last year with a record of 30,347 numbered bulls, and a proportionate number of cows. More than 700 subscribers and breeders are enumerated in the last volume.

The value attached to the best specimens has of late years become almost fabulous. The late Mr. Whitaker, whose sound judgment and modest opinions entitle his observations to the most sincere respect, says in 1829 that 2000 guineas would not purchase ten of the most select animals in the country. That price now breeders have been known to steadily and calmly decline for one animal. The cattle plague of 1855-6, followed by three years of drought, has also affected prices. Farming has become a more fashionable pursuit, and many leading men of the country have stocked their own estates with pure-bred cattle, while the retired merchant has amused his leisure hours with model farming, and found a pleasing relaxation in breeding Shorthorns. The demand has consequently exceeded the supply; prices have risen, and breeders of the more fashionable tribes have endeavoured to keep up the supply by offering portions of their herds, which have sold at enormous averages. Thus we find pure specimens of bulls and heifers, making, even at auctions, 1000 guineas each, and private purchasers have gone as high as 1500 guineas for heifers. The summary of the principal sales held during the year 1871 shows that about 2064 pure-bred Shorthorns of all ages have been sold for 115,400*l.*, averaging within a trifle of 56*l.* each.

[Since the above was written, all previous results have been eclipsed by the memorable sale of Mr. Campbell's Duchess and Oxford cattle, at New York Mills, in America, when eleven females of the Duchess tribe averaged 4522*l.* 14*s.* 2*d.*, the highest price being 40,600 dollars paid for 8th Duchess of Geneva, which was bought for Mr. R. Pavin Davis, of Gloucestershire, contrary to that gentleman's orders, and resold by him to an American, and has since died. Lords Bective and Skelmersdale and Mr. Holford were the other English buyers. Three Duchess bulls averaged 1638*l.* 15*s.*; Second Duke of Oneida, a remarkably fine animal, making 12,000 dollars. With one exception the Oxfords were purchased by Americans. Six females averaged 1087*l.* 10*s.*, and the bull calves 396*l.* 16*s.* 8*d.*—ED.]

The preponderance of the breed at the meetings of the Royal Agricultural Society of England has been remarkable. The result of seven years, ending in 1852, was 702 Shorthorns against 211 Herefords and 357 Devons; and for the last ten years the numbers exhibited have been 1476 shorthorns, 574 Herefords, 472 Devons. At the leading markets and fairs, except, perhaps, in the south-west, they comprise the majority; and it is estimated that there are more Shorthorns bred, fed, and grazed in England than all the other breeds put together.

The great milking properties of this breed have made them equally serviceable to dairymen and graziers, and one of the more recent companies in London started under the name of the Royal Shorthorn Dairy. Years ago it was customary to walk the ordinary unimproved Shorthorn from the north to the south of England; farmers would meet the droves on the road, buy the best animals, and in this way many capital stocks have been established in the midland and southern counties. Of late years complaints have been frequently made that Shorthorns are not good milkers. This has doubtless been correct in many instances, and has risen, not from inherent defects, but from the pernicious effects of forcing young animals into a condition of premature fatness. There are, however, animals of all strains, capable not only of making their own calves fat, but of giving several quarts of milk daily in addition.

One great advantage of the Shorthorn is its marvellous efficacy in crossing and improving other breeds. In Scotland many of the native black herds have been crossed generation after generation, until the characteristics of the Shorthorn are remarkable. In Wales, the "coloury beast," as the Shorthorn is called, is gradually working upon the Castlemartins and runts, and pure-bred herds are to be found in the south as well as the north of the Principality. The marvellous improvement in the Irish cattle has become proverbial, and so recently as last year a breeder in county Wexford gave 750 guineas for a yearling heifer. Even the Isle of Man boasts its pure herds and a 400-guinea heifer; whilst the Orkneys and Shetlands are not destitute of pedigreed bulls.

But it is to the New World that the greatest importations have been made. America imported pure Shorthorns upwards of fifty years ago, and every year numbers of cattle leave our shores for the States. Spirit and enterprise have been rewarded, and the offspring of animals imported a generation back have of late years found their way back to our own herds. The vast area and rolling plains of the Western States are affording fine fields for grazing and breeding, and what has for years been done in Australia, where numberless bulls and also heifers have been sent, is now being practised in the Far West. Canada, too, with its fields five months white with snow, finds the purest pay the best, and one energetic Canadian in 1870 spent 20,000*l*. in importing pure animals, which have been highly remunerative.

New Zealand has also its breeders and importers; and, coming nearer home, we find France took thirty years ago some of our best cattle, and one of her first acts after the late terrible calamity was the importation of four pure-bred Shorthorn bulls. In Germany several pure herds are to be found, and Shorthorn bulls have been used among the native breeds of Russia and Bessarabia, and even in Egypt. It is in the prepotent powers of this impressive race that its great value lies. Its adaptability to all climates and soils, its marvellous faculty of growing and fattening at the same time, its maturity at an age when other cattle are considered but half-grown, its faculty of raising its own offspring with a bountiful supply of milk, insure its great and permanent superiority.

The Shorthorn has been called the "Universal Intruder." Wherever Britons colonise, the Shorthorn makes his home. In many a distant land, where the English tongue is unknown, his influence is extending, and he undoubtedly is the great means of transmitting to other times and other nations that great national institution, the "Roast Beef of Old England."

CHAPTER VII.

HEREFORD BREED OF CATTLE.

BY T. DUCKHAM,
EDITOR OF THE "HEREFORD HERD BOOK."

THE HEREFORDS are an aboriginal race of cattle indigenous to the soil of the county from whence they take their name. Yet experience has proved that the exercise of sound judgment in making selections for breeding purposes is alone requisite to insure the success of those who breed them in almost every known climate. They are of the middle-horn tribe, and have for ages past been highly esteemed for their fine quality of flesh, which, by the intermixture of fat and lean, presents that marbled appearance so much prized by the epicure, and commands a top price in the market. The rapidity with which they lay on fat is certainly unsurpassed, if equalled. Experimental trials have been made with them and selected specimens of other pure breeds, which have ended in the uniform result that they yield the best return to the grazier for the food consumed. The value of the cattle of the district has been noticed by different writers for many centuries past. Speed, in his history of the county, says: "The county's climate is most healthful and temperate, and soyle so fertile for corn and cattle that no place in England yieldeth more or better conditioned."

The principal herds are in the hands of the tenant farmers of Hereford and adjoining counties, and have been handed down generation after generation from father to son in all their purity. The steers are looked upon as the rent-payers of the district, and perhaps no finer sight of cattle can be seen in the kingdom than that of the Hereford October fair, where several thousands line the streets of the ancient city, and by their distinctive marks and uniform appearance lay claim to each other as kindred of the same family. They have long been sought after by the graziers of the Midland and Eastern Counties. Amongst those of bygone days the name of Westcar, of Creslow, Bucks (one of the most active founders of the Smithfield Club), stands pre-eminent. His forty years' attendance at the Hereford fairs was commenced in 1799, and his twenty first prizes in succession at the Smithfield Show was no mean achievement. The indefatigable hon. sec. to the Smithfield Club, Mr. B. T. Brandreth Gibbs, in his tabulated statement of the prizes awarded, has shown that during the first fifty-two annual meetings of the club, when all breeds met in competition with each other, 185 prizes were awarded to Hereford steers or oxen, whilst only 190 fell to the lot of all the other breeds or cross-breeds put together.

The production of steers to meet the demand of the graziers being the chief aim of the breeders, and the well-known influence of the male animal for breeding purposes, little attention has been paid to the cow. It has been thought sufficient for her to possess the qualifications which long experience has proved to be necessary to insure success with her progeny; her milking properties have been much neglected, and the calf usually allowed to run with its dam. Owing to the almost uniform adoption of this system, she has obtained the character of being a bad milker; but in other districts, where the milking properties are cultivated, it is not so, and, as her aptitude to fatten surpasses that of most other breeds, and she consumes less food in proportion to the quantity of meat made, she is gaining favour in many of the dairy farms of Dorset, Gloucester, Somerset, Cornwall, &c.

Duncombe, in his "Farming of Herefordshire," published in 1813, says: "Large size, an athletic form, and unusual neatness, characterise the true sort; the prevailing colour is a reddish brown, with white face." But in 1845, when Mr. Eyton issued his first volume of the "Herd Book," he found it requisite to divide them into four classes—viz., mottle-faced, light grey, dark grey, and red with white face. As each of the three first-named classes are now nearly extinct, I will briefly notice the characteristics of the latter.

The face, throat, chest, lower part of the body and legs, together with the crest or mane, and the tip of the tail, a beautifully clear white. The horns yellow or white waxy appearance, frequently darker at the end; those of the bull should spring out in a nearly straight line from a broad, flat forehead; whilst those of the cow have a wave or slightly upward tendency. The countenance, pleasant, cheerful and open, presents a placid appearance, denoting good temper and that quietude so essential to the successful grazing of all ruminating animals; yet the eye is full and lively. The head is small in comparison to the substance of the body; muzzle white, and moderately fine; cheek thin; chest deep and full; shoulder blades thin, flat, and sloping towards the chine, and well covered on the outside with mellow flesh; kernel well up from the shoulder point to the throat, and so beautifully do the blades bend into the body that in a first-class, well-fed animal it is difficult to tell where they are set on; the chine and loin broad; legs straight and small, the rump forming a straight line with the back; thighs full of flesh to the hocks; a well-sprung rib and deep flank. The whole carcass well and evenly covered with rich mellow flesh, distinguishable by its yielding with a pleasant elasticity to the touch, and a hide thick yet mellow, well covered with soft glossy hair, having a tendency to curl.

The "Herd Book" was commenced by Mr. T. C. Eyton, of Eyton Hall, Salop, in 1846, and the two volumes published by him contain the pedigrees of 901 bulls, but no cows. A second edition of those volumes, and five other volumes, have been published by the writer. The work now contains the pedigrees of 3636 bulls, 3968 cows, and 3540 heifers, and notices of the various prizes won by the animals entered; it is handsomely embellished with faithful likenesses of first-class breeding animals—the qualification for their being so placed is their having won a first prize at a show of the Royal Agricultural Society of England. To the "Herd Book" is now added, as a companion work, a "Record of Hereford Transactions," to which for the future all notice of prize animals will be confined, with the exception of those lithographed to embellish the "Herd Book." This record contains reports of sales by public auction and private contract, exportations, and other matters of interest to breeders. It is published in half-yearly parts.

Sixty years have passed since Mr. Grove first introduced the Herefords into Dorsetshire, and many valuable herds are now established in that county. The late Earl of St. Germains gave them a place on the west banks of the Tamar some half-century or so ago, and several good herds now graze the pastures of Cornwall. They are fast supplanting the native breeds of the counties of Glamorgan, Brecon, Radnor, and Montgomery. The Earl of Lisburne was foremost in placing them on the mountain sides in Cardiganshire, and the readiness with which they became acclimatised was such that his lordship's tenantry and others gladly availed themselves of the use of the bulls at Crosswood Park with the little black cows of the district. The result was most satisfactory, and the influence of the pure-bred sire is such that it is to be seen for several generations. The cross-breds are equally hardy as the natives; they feed more kindly, attain greater weight, and are more prized by the butchers. The late Mr. Lumsden, Auchry House, Aberdeenshire, first introduced Herefords into Scotland. Amongst his early purchases were some cows in the Hereford Fair and a 100-guinea bull, Matchless (415), at Mr. Hewer's sale in 1839. Mr. Lumsden made many trials with pure-bred Herefords, Shorthorns, and Aberdeens, and their various crosses, but always maintained the result was in favour of the "red and white faces." Mr. Copland Mill, of Ardlethen, Aberdeen, writes: "In regard to my experience of crossing with a pure-bred Hereford bull, I have now done so with from forty to fifty cows for the past seven years, and have no cause to regret it; for I find that my cattle are improved in weight, that they come sooner to maturity, and that their constitution is very much improved. So much is this the case, that a good many of my neighours who were prejudiced against the Herefords are now breeding from them." The Earl of Southesk has established a choice herd in Forfarshire, and I am informed that "the whole herd present a kindly appearance. A demand has been created for bulls at good prices, and those which have been sold have given great satisfaction. The generally acknowledged hardiness of constitution and fine character is well maintained and transmitted to their offspring." The late Prince Consort—that noble patron of all that was good and virtuous, whose desire for this country's greatness induced him, amongst his other pursuits, to become a pioneer in agricultural advancement

—laid the foundation of the royal herd of Herefords at the Flemish Farm, Windsor, in 1855, and the marked success which has for several years past attended the exhibits from that herd at the various national shows is the best proof of the correctness of the judgment displayed in the selections. Mr. R. W. Reynell, of Killynon, Westmeath, Ireland, whose herd was established a century or so ago, says that he has fed them with other pure breeds, and contends that the Herefords are the fastest feeders he knows, particularly on grass. Mr. P. J. Kearney is equally satisfied with the doings of his herd at Clonmel; and so is Mr. Gilliland at Londonderry, where it has been my pleasure to see them retaining their characteristics in all their excellence.

Thus I have shown the satisfaction which the breeders of Herefords experience in their use from the extreme west and south of England to the north of Scotland, and from the south to the north of Ireland.

They were introduced in Jamaica by Mr. Malcolm in 1845, and in Trollope's "Travels in Jamaica" he says, "At Knockalva I looked at Hereford cattle, which I have rarely, if ever, seen beaten at any agricultural shows in England." Mr. J. Edwards, the manager, writes: "The cross with the Hereford bull and native cow is so direct that the bull carries all before him, and many of our half-bred cattle you would scarcely suspect as being any other than pure breds. Here we require a breed of cattle to be good workers, hardy, and of great aptitude to fatten, and I fear no contradiction when I say that no breed displays those qualifications in so eminent a degree as the Herefords."

Mr. F. W. Stone, of Moreton Lodge, Guelph, Canada West, has for many years been an extensive and successful breeder of Shorthorns; but, in addition to those, he, ten years ago, resolved to establish a herd of Herefords, and he writes that he believes the Hereford preferable to other breeds as grazers. Many valuable Herefords bred by him have travelled far away into the United States, and in various exhibitions, when competing with other breeds, have carried all before them. That has been particularly the case with Mr. H. C. Burleigh's selections; 700 dollars, four silver medals, and twenty-seven diplomas represent his two year's prizes.

There are several herds in the United States, and perhaps the most extensive are in the hands of Mr. J. Merryman, Treasurer of the State of Maryland; Mr. G. Clarke, Springfield, Otsego County, N.Y.; Mr. W. W. Crapo, Flint, Michigan; Mr. H. C. Burleigh, Fairfield, Maine; and Mr. W. W. Aldrich, Elyria, Ohio. Mr. Sandford Howard, Secretary to the Michigan Board of Agriculture, in his report in 1868, gave a letter from the late Governor Crapo, in which he says: "The Herefords have done extremely well. They have no more than ordinary fair keeping, yet they are in prime condition. I have little doubt that the Herefords will yet be the stock for Michigan. They are docile and hardy, besides being very easy keepers, and I have no doubt will stand a long, severe winter, and come out ahead of the Shorthorns in the spring, on two-thirds the cost of keeping. I intend, however, to give the Herefords, Shorthorns, and Devons a fair trial, both as full bloods and grades." Mr. W. W. Crapo has written me: "My father, the late Governor Crapo, of Michigan, and myself, have been for several years engaged in the breeding of Hereford, Shorthorn, and Devon cattle, having in view the testing of their relative merits. The result thus far is decidedly favourable to the Herefords.

Mr. E. Maclean, Butley Manor, Auckland, New Zealand, purchased a lot of Hereford cows in Australia, and bulls from Her Majesty's herd. He writes: "I now breed about 100 calves a year, mostly Herefords, as I am gradually weeding out all others. I have bred Devons and Shorthorns mostly, the latter for the past twenty years in this province, but I much prefer Herefords."

Many large and valuable herds are found in Australia, and numerous importations from several of the best herds in England have been made during the past few years by Mr. J. White Martindale, Hunter's River; Mr. J. Nowlan, Eelah, West Maitland; Messrs. Barnes and Smith, Dyraaba, Richmond River; Messrs. Livingston, Learmonth, Ercildoun, Victoria; Mr. Angas, Angaston, South Australia; Mr. Robertson, Lake Colac, Melbourne; Mr. A. Bloxsome, Rangers' Valley, New England; Messrs. Mort and Co., and Messrs. Dangar and Co., Sidney; Mr. J. Price, Hindmarsh Island, South Australia; Mr. G. Loder, Singleton; Messrs. W. and F. Fanning, of Wooroowalgan, Richmond River, &c.

The *Sydney Morning Herald*, Aug. 12, 1870, in its report of the Singleton Show, says: "A glance at the cattle pens could not fail to establish incontestably one fact of great importance to the cattle breeders of this district, viz., the marked superiority of the Herefords as contrasted with the Durhams (Shorthorns). They were not only most numerous, but in better order, showed better breeding, and in every way superior to the shorthorns. There is no mistaking a pure-bred Hereford." The same paper (Sept. 7, 1870), in an able article on the Agricultural Resources of New South Wales, says: "The debate concerning the merits of Shorthorns or Herefords is very strong. Both breeds have many advocates. It is generally admitted, however, that the Hereford travels better than the Shorthorn, and better endures periods of dearth and drought. A vast quantity of cattle of this colony having to travel 500 to 800 miles to the slaughter-house, this quality is a consideration of the utmost consequence."

The *South Australian Register*, Oct. 11, 1870, in the report of the Adelaide meeting, says: "The first place in the programme is conceded to Mr. Price's imported Hereford bull. Never was priority of position more deserved." He was acknowledged to be the monarch of the yard. The *Maitland Mercury*, Aug. 17, 1871, in the report of the Singleton Show, says: "Such beasts as were exhibited are seldom seen in this or any other district; it was impossible to view them without thinking of the roast beef of old England," and thus describes Mr. White's prize bull Prince of Wales:—"With a coat of velvet, a mild, gentle eye, and quiet temper, that seems more like that of a lamb than that of a bull." And Mr. A. A. Dangar's colonial-bred heifer: "We think, in quality and condition, this animal is perfect." The *Australasian*, in the report of the Sidney Show, 1871, says: "The feature of the exhibition was undoubtedly the cattle, and, of the cattle, especially the Herefords. Without disparaging the exhibition of Shorthorns, it is but fair to state that they did not equal the Herefords; nor, if the exhibition of fat stock be any criterion of the success of the breeders, did they prove any essential adaptability to either climate or pasturage. Mr. White's pen of fat oxen were magnificent animals, and they appeared to assert their supremacy as being the primest of the prime; and the Hereford cattle will remain on the annals of the exhibition of 1871 as the main feature of excellence." Hereford bulls have recently been exported to South America, and their value as a cross upon the native breeds is being tested.

CHAPTER VIII.

DEVONS.

By CAPTAIN J. T. DAVY,
EDITOR OF THE "DEVON HERD BOOK."

THE BREEDS OF CATTLE reared on farms are very numerous, and often approximate one to another by a series of the nicest and almost imperceptible gradations. Where a breed has found a congenial soil and climate, it seems to flourish almost in spite of neglect. The early history of cattle speaks of three kinds, viz., the long-horned, found in the midland counties and in Ireland, the short-horned, in the eastern and northern counties; and the middle-horned, in the western part of England, in Sussex, and in Scotland.

The Devon breed belongs to the middle-horned variety, is evidently an aboriginal one, and there is little or no doubt that Devon, Hereford, and Sussex cattle, and probably also those of Wales and Scotland, were originally descended from the same stock. They have all the characteristics of the same breed, changed by soil, climate, time, and by being subject to man's will and control. These influences change the capabilities and characteristics of most breeds of animals coming under the denomination of stock. The late Mr. Youatt, in his valuable work on cattle, speaking of the skulls found in different parts of England, says: "There is a fine specimen in the British Museum; the peculiarity of the horns will be observed resembling smaller ones dug up in the mines of Cornwall, and preserved in some degree in the wild cattle in Chillingham Park, and not quite lost in our native breeds of Devon and Sussex, and those of the Welsh mountains and the Highlands."

The middle-horned varieties are fairly good milkers, but are remarkable for the *quality* rather than for the *quantity* of their milk, which yields a large proportion of cream and butter. As a general rule, the better the milking properties of cattle, the more are they disposed to internal accumulation of fat; and it should be understood that excessive accumulation of this kind are the farmer's loss and the butcher's gain.

It is not difficult to observe that the Devonshire and Sussex races are of the same extraction; so nearly do they resemble each other in colour and length of horn, that if any one unacquainted with the distinctive features of those two breeds were shown two animals, one a Devon, the other a Sussex, he would find it difficult to detect any material difference between them, except that the Sussex beast might appear rather the larger or the "taller," from the greater length of leg. A more experienced observer would notice in him a less finely chiselled head, coarser eyelids combined with a less pleasing expression of the eye, and a crescent-shaped, upward horn, instead of the deer-like head and gracefully curved waxy-looking horn of the Devon. In their shape and size, as well as in the curve of the horn and the heavy eye, the Sussex cattle bear a strong resemblance to those formerly bred about Taunton, before the latter were so much mixed with Hundred Guinea (56) and other North Devon bulls.

Particular breeds and their varieties were formed long before the modern scientific system of breeding was established. We find that large breeds and bulky varieties of the same are co-extensive with a warm climate and rich herbage, and that smaller breeds and their varieties pervade those districts where the pasturage is more scanty and the climate colder. To wit, we find a larger variety of Devons with long straight hair bred in the fertile vale of Taunton Dene; while the Devon reared in North Devon is noted for his soft, rich, curly coat of hair, which he frequently loses when taken into other and richer districts. It is well known that the general appearance and hair of the bull Hundred Guinea (56), who was purchased by Messrs. Bult and Bond, near Taunton, of the breeder, at Molland in North Devon, at the foot of the Exmoor hills, altered considerably after two years' residence in his new home. The effect of soil, climate, and water on the colour and hair, and in developing changes in the form and physical structure, is well known and duly appreciated; so much so, that a great number of animals are sent to North Devon for summering, with the twofold object of grazing and changing their coats.

[As an illustration of the influence of climate and food, we may refer our readers to the difference between the ordinary Devon, and the South Hams cattle, a splendid collection of which were exhibited in competition for the Devonshire Agricultural Association prizes, at the Plymouth meeting of the Bath and West of England Society, in June 1873. These animals are much larger and coarser in bone, and closely resemble the Sussex cattle, though hardly exhibiting so much quality; they are hardy, milk well, and are esteemed in their district, which comprises the country on the right of the railway from Plymouth to Totnes; they require more time to mature than the more compact North Devon, and would hunger where they would thrive. Amongst others we may name Mr. William Coakes, of Charleton Court, Kingsbridge, as a highly successful breeder.—ED.]

Notwithstanding his curly hair, the skin of a Devon must be mellow and elastic. Experience shows that some animals fatten faster than others. On "handling" them, we find the skin and parts beneath soft and "mellow." This "mellowness" is a kind of softness or elasticity perceived upon pressing the skin with the fingers, and is a favourable sign of the aptitude of an animal to fatten.

These parts are the cellular membranes, which in fat animals are full of fat, and the possession of this mellow feeling by store stock denotes that there are plenty of membranous cells ready for the reception of fat. None have been more thoroughly successful than the Devon breeders in attaining this desirable object; they consider an animal of little value if it cannot be fattened without very extraordinary food.

The general form of a Devon is very graceful, and exhibits a refined organisation of animal qualities unsurpassed by any breed. The expression of the face is gentle and intelligent; the head small, with a broad, indented forehead, tapering considerably towards the nostrils; the nose of a creamy white; the eye bright and prominent, encircled by an orange-coloured ring; the jaws clean, and free from flesh; the ears thin.

The horns of the female are long and spreading, gracefully turned upwards, and tapering off towards the ends. The general aspect of the head should in many points resemble that of the deer. The horns of the bull are thicker set and more highly curved, in some instances standing out nearly square, with only a slight inclination upwards.

Red is the true Devon colour, which varies from a dark to a lighter, or almost to a chesnut shade. In summer the skin is mottled with beautiful spots of a slightly darker shade than the ground colour of the skin.

The outline of a fat Devon very nearly approaches a parallelogram. The frame is level from the tops of the shoulders to the tail; the belly is longitudinally straight, and well filled out at the flanks. The breast is wide, coming out prominently between the fore legs, and extending downwards almost to the knee joint. The neck is long and thin, increasing towards the shoulder, which is tapered off to meet it. The ribs project at right angles to the back, with wide flat loins, and long rumps well filled out, thus enabling them to be loaded with more beef in the most valuable parts than almost any other breed.

As converters of vegetable into animal food, breed against breed, they return as much per acre, or for weight of food consumed, as any. Animals possess no magical power of producing beef, except from the food which they consume; it is therefore contended that, if the herbage of any given number of acres were to be consumed by Devons, they would produce in the aggregate as much beef as any other breed, a greater number being required to consume it; at the same time there would be a greater weight of the most valuable beef, and less of the coarse joints and offal. This is the reason why Devons and Scotch cattle sell first in the morning, and command the best prices, in the London and other markets. Mr. Wainwright, a Devon breeder in the State of New York, says: "Their beef is of a fine quality, and brings a high price in the markets. They withstand extremes of temperature. On a poor pasture, from their peculiar build, they are enabled to travel rapidly over the ground without fatigue, and get sufficient nourishment where a heavy Shorthorn or Hereford would starve. The very best of this breed are the best in the world." Mr. Steinmetz, of Pennsylvania, writes me as Editor of the "Devon Herd Book";

"I find North Devon Cattle the most profitable breed in America; I can raise more valuable beef on them with the same amount of food than any other breed."

What is meant by a gold medal beast at our shows? That animal which most nearly approaches to the form and quality of North Devon; it is the length, depth, and width, not the height of a beast, which constitutes size. The cry has been for the animal that will be the first ready for the butcher, and the Devons have answered it. They bear the change of soil and climate well, *thrive* where many breeds would *starve*, and rapidly outstrip most others when they have plenty of good pastures. That they are a good rent-paying breed, especially in cold, hilly districts, is clearly proved by the fact that the majority of the oldest and most successful breeders are tenant farmers, whose ancestors have kept them for the last 150 or 200 years, in most cases on the same farms, in North Devon and West Somerset, frequently at an elevation of 600 or 800 feet above the sea level.

A few years ago, the late Prince Consort established a herd of Devons, and they are also patronised by several noblemen and gentlemen in various parts of the United Kingdom. They have also been conveyed to new homes in the United States of America, where there are a great many herds of the purest descent; and to Australia, Natal, Mexico, Jamaica, Canada, and France, in all of which places they are answering a good purpose. "The Devon Herd Book" was first compiled by the writer and published in 1851, the 5th volume bringing up the number of bulls to 977, and females to 3143, was published in 1869.

Every breed and its varieties possesses peculiar merit, each answering a better purpose than the other, according to the soil, situation, and other circumstances in which it may be placed. To succeed, we must study to keep animals which are suited to our soil, pasturage, and climate. Those animals which will thrive in cold, bleak, hilly districts cannot fail to flourish in more favoured situations; and North Devons are never seen to greater perfection than among their native hills, the last haunt in England of the wild red deer, and where

At morn the blackcock trims his jetty wing.

[It is hardly doing full justice to the Devon to conclude a notice, however short, without reference to their superior qualities for draught purposes. It is true, that at the present time, working cattle are the exception. The value of meat and the advantage from getting our animals early into market having operated against the practice, but no one can say how soon circumstances may arise which will render it desirable to recur to bullock labour. The extreme price of horse flesh lately may well cause us to reconsider the question, and if, as is possible, the supply of cattle either from increased breeding or foreign sources should exceed the demand, a part of our surplus stock may advantageously be employed in this way. Without disparaging other breeds, we are bound to state that the North Devon one is quite unrivalled as a worker, and this is due to his activity and strength. We have not a ponderous over-weighted animal, good at a dead pull no doubt, but hardly able to crawl under its own weight, but we have a class of cattle that with proper training are capable of walking as fast and getting through as much work as heavy draft horses. The late Mr. Plumb, of Ashton Keynes, near Cirencester, whose land was of a light sandy nature, worked Devon oxen in pairs, and his teams could hold their own at the ploughing matches, the work being done quite as expeditiously as where horses were used. These bullocks were yoked with collars and guided by reins. When a dead pull was required, or when a load of sheaves got fast in the fold yard, a bullock was much more efficacious than the strongest horse. Mr. Plumb knew the value of good food, and whilst he worked hard he fed well. These bullocks on an average consume a bushel of corn a week, with plenty of bulky food, his practice was to buy in each year three-year-olds, work for two years, and send off fat when turned five years old. Another point that may be urged in favour of the breed is that though careful breeding and selection has only been practised of late years, they have always been noted for symmetry and quality.—ED.]

CHAPTER IX.

THE LONGHORNS.

THE PRESENT POSITION OF THE LONGHORN breed of cattle illustrates the old saying that " every dog has its day." Confined now to a few amateur farmers in the midland counties, it is difficult to realise that a hundred years ago they were the most valuable breed in this country; yet such is the fact. Yorkshire has the credit of giving rise to the Longhorn and their supplanters, the Shorthorn. The latter, however, originated in the eastern division, whilst the district of Craven (the original home of the Longhorns) is in the West Riding, bordering on Lancashire, from whence they spread out into the latter county and the south-eastern portions of Westmorland. Like the Durham cattle, they enjoyed a considerable local reputation, those bred in the fertile vale of Craven being considered the quickest feeders, as they were the handsomer beasts; but it required the genius of Bakewell to draw them from their comparative obscurity, and give them a reputation which at that time seemed unassailable. Culley states that before Bakewell's time "The kind of cattle most esteemed were the large, long-bodied, big-boned, coarse, flat-sided kind, and often lyery or black-fleshed." This, however, is rather a sweeping denunciation, and, though applicable enough to the general run of Longhorns as they appeared in the various counties to which they had gradually spread, either tolerably pure or incorporated with the prevailing breeds of the district, must not be taken as a fair account of the Craven cattle, many of which were noticeable for rotundity, length of carcase, mellowness of skin, and quality of their milk. The improvement of the breed dates from 1720. At that time a Sir Thomas Gresley had a choice selection at Drakelow House, near Burton. A blacksmith and farrier of the name of Welby, who resided at Linton, in Derbyshire, on the borders of Leicestershire, purchased some valuable animals from Drakelow House, and took much pride in improving the stock. After carrying on for a few years with manifest success, a disease broke out and carried off the greater part of the herd. Mr. Webster, of Canley, near Coventry, is the next name of note in Longhorn history. How comparatively unknown now are those who worked for their sort as successfully as Collins, Mason, Bates, &c., did for the Shorthorns. The latter have become familiar in our mouths as household words, as helping to illustrate the rise and progress of a breed of world-wide celebrity, whereas the pioneers of the Longhorns are known only to the student—*sic transit gloria mundi*. Mr. Webster also worked upon Sir Thomas Gresley's stock, using bulls from Lancashire and Westmoreland. He bred a celebrated bull named Bloxedge, which produced some remarkable stock. It is unfortunate for posterity that Bakewell was not large-minded enough to leave us a record of his work. It would be of great interest as well as advantage to know how he set to work to develop the improved Leicesters, as his Longhorns were soon christened and known for many years. We, reasoning by analogy, can only surmise that he went for quality rather than size, and, as in his sheep, strove after correct outlines, fine bone and offal, with great aptitude for feeding. This last quality he paid most attention to, and naturally sacrificed to it other points, more especially hardiness and yield of milk. Bakewell found that feeding properties were to a great extent hereditary and could be perpetuated by close breeding; he, therefore, preferred improving his stock by selecting animals of the same kind rather than run the risk of crossing. He commenced his cattle breeding with two heifers from Canley, using on them a Westmoreland bull; and as far as is known he never went further, at first breeding very closely, but as the herd increased he was able to unite more distant affinities. In a few years his stock became known for rotundity of outline and aptitude to feed. They were much prized for feeding, but did not fill the pail like the old sort. Twopenny, out of one of the Canley heifers, by the Westmoreland bull, was very celebrated. His offspring, a bull named D., was even more remarkable; he was very closely bred, being by a son of Twopenny out of a daughter and sister of the same bull. Following in the steps of Mr. Bakewell, came Mr. Fowler, who farmed in Oxfordshire. His cows were of Canley breed; whilst his bull, Shakespeare, considered the best he ever bred, was by D. out of a daughter of Twopenny. Mr. Marshall, in his "Economy of the Midland Counties," gives a good description of this bull, which, save in his horns, did not resemble the Longhorns so much as the Durham of that day: "His head, chap, and neck remarkably fine and clean; his chest extraordinarily deep; his brisket down to the knees; his chin thin, and rising above the shoulder points, leaving a hollow on each side behind them; his loin narrow at the chine, but remarkably wide at the hips, which protruded in a singular manner; his quarters long in reality, but appearing short, occasioned by a singular formation of the rump. At first sight it appears as if the tail, which stands forward, had been severed from the vertebræ by the chop of a cleaver, one of the vertebræ extracted and the tail forced up to make good the joint; an appearance which, on examining, is occasioned by some remarkable wreaths of fat formed round the setting on of the tail—a circumstance which in a picture would be deemed a deformity, but as a point is in the highest estimation. The round bones snug, but the thighs rather full, and remarkably let down. The legs short, and their bone fine. The carcase throughout (the chine excepted) large, roomy, deep, and well spread." The value of this bull was fully appreciated by Mr. Fowler, who, except letting him for two seasons to a Mr. Princess at 80 guineas a season, retained him for his own use. In the year 1791 Mr. Fowler had a public sale, at which fifty head of cattle produced 4289*l.* 4*s.* 6*d.*, being an average of over 80*l.* a head, prices that would be regarded as satisfactory for a choice selection of shorthorns at the present time, and which show the high estimation in which the improved Leicesters were held, and the advance made by such men as Bakewell, Prinsep, and Fowler. Mr. Marshall thus describes the character of the improved Leicesters: " Forend long and light (this we may observe is a fault apparent both in the few herds remaining in this country and in the Irish imports, unmistakably of Longhorn origin, of which more anon); neck thin, head fine but long and tapering, eye large, bright, and prominent. The horns vary with the sex; those of bulls comparatively short, from 15in. to 2ft. The oxen extremely large, from 2½ft. to 3½ft. Cows nearly as long, but fine and more tapering. Most of the horns hang downward by the side of the cheeks, and then if well turned, as in many of the cows, shoot forward at the points; the shoulders fine, thin, and well placed—this was particularly noticeable in the Dishley herd—girth small, as compared with shorthorn and middlehorn breeds; the chine remarkably full when fat, but hollow when low in condition; loin broad, and hips wide and protuberant, the quarter long and level, fleshy thighs, with small clean, but comparatively long legs; carcase round, and ribs well sprung, flesh of good quality, hide of medium thickness, and colour various—the brindle, the finch-back, and the pye most common. As grazier's stock, they undoubtedly rank high; the bone and offal small, and the forend light, while the chine, the loin, the rump, and the ribs are heavily loaded, and with flesh of the finest quality. In point of early maturity they have also materially gained; in general they have gained a year in preparation for the butcher." Such was the character of the improved Longhorn as established by these leading breeders. With such excellence, how is it they so soon disappeared from this prominent position? Possibly those who followed were not able to maintain the character and quality of the stock; but more probably the increasing popularity of the Durham or Shorthorns caused them to be shelved; and it is a noteworthy fact that, at the present day, as far as we know, Leicestershire does not possess more than one herd of the old sort, whereas in their original homes, viz., in Westmoreland, Lancashire, and Yorkshire the Shorthorn is dominant. In Ireland the Longhorn influence was undoubted; whether this was due to importation of English cattle, or *vice versâ*, we are unable to say, as records are wanting. Anyhow, they were at one period the prevailing breed in a large portion of Munster; and we can trace their influence in the cattle from that district to this day. Shorthorns, however, are rapidly becoming predominant through the length and breadth of the Emerald Isle, and to them is due in great measure the marvellous improvement of late years which is rapidly placing the Irish on a footing with the home-bred.

Sixty years ago the Longhorn was the most important and fashionable breed of cattle inhabiting the counties of Durham and Stafford, and there still lingers in the district wondrous tales of the quantity of milk yielded by some favourite cow, or the more marvellous weights which the oxen and heifers attained when grazed on the rich alluvial pastures of the Trent, the Dove, or the Derwent. Viewed by their narrators through the mists of a long series of bygone years, their merits became magnified to a degree sufficient to awaken in our minds a feeling bordering on incredulity. It will give the reader some idea of the state of agriculture in this district when we state that down to about the period to which we refer the cattle were all wintered in the fields, and within five minutes' walk of where we write stands the first cowshed ever erected in this parish, or within a radius of several miles. The farmers attending the markets of Derby and Loughboro' would no doubt freely discuss all the floating rumours concerning the doings of their neighbour of Dishley. With such associations, the breed could not fail to attain a high repute. Down to within a recent period the late Mr. Bakewell, a descendant of the celebrated breeder of that ilk, kept a fine herd of Longhorns on his farm at Lockington, on the banks of the Trent. Amongst the general stock of the district are still to be met with many Longhorn crosses, and within the last three years we have seen large dairy farmers use a Longhorn bull on their common Shorthorn cows. Long before the days of root cultivation, and prior to the use of artificial food or stall feeding was generally practised, we have it on authentic authority that Christmas oxen which had never known the shelter of a roof-tree, and whose only food from weaning time consisted exclusively of hay and grass, would frequently reach the great weight of 360lb. a quarter. To attain this was a sufficient proof of the merit of the breed. From some cause or other the breed has gradually receded in public estimation, the only herds of note now existing in England are those of his Grace the Duke of Buckingingham, at Stowe Park, Bucks; Sir John Harper Crew, Bart., Calke Abbey, Derbyshire; John Godfrey, Wigston Parva, Hickley, Leicestershire; and R. H. Chapman, of Upton, Nuneaton, Warwickshire. The Duke of Buckingham has a herd of fine animals, numbering at the present time nearly 100 head; they are directly descended from the Bakewell, Canley, Rollright, and some of the purest old Warwickshire families. The herd is of long standing, and has been bred with great care and judgment. Animals from this herd have frequently distinguished themselves in the show yard. Sir John Harper Crew, who was formerly an admirer and breeder of pedigree Shorthorns, and, though it is scarcely ten years since he commenced to cultivate the Longhorn, now possesses a herd of forty breeding cows, which for uniformity of type could scarcely be excelled by their great rivals, the Shorthorns. They come to hand mellow to the touch. The skin, though thick, is covered with a profusion of rich, soft hair; the rib is well sprung, chine broad, shoulders well placed, barrel round and deep, the general appearance in unison, denoting a healthy and vigorous constitution. They are good milkers, and, as a rule, prolific breeders. What appears to us as their only weak point in these days of high feeding and quick returns is that they are longer in arriving at maturity than the improved Shorthorn; consequently they give a less return for the quantity of food they consume. T. W. Cox., Esq., of Spondon Hall, near Derby, has a very good herd, tracing their account from the herd of Bakewell and the old Warwickshire sorts, on an outlying farm which he has in hand. Both rearing and cheese-making are practised at Spondon. A small herd is kept for the use of the hall. All the steers are made off to the butcher at about three years and a half old. Mr. Cox has long been a successful exhibitor at the Royal, as well as the Smithfield and Birmingham fat shows, and also at many of the local societies in the neighbourhood. The family of Cox have been celebrated for many generations as breeders of Longhorns. Mr. Chapman, of Nuneaton, keeps a herd of thirty cows; they are heavy-fleshed animals, of the old Warwickshire blood. All the calves are reared; the most promising males are saved as bulls. Cheesemaking is practised, and Mr. Chapman expresses himself well satisfied with the quantity of the produce; the milk is considered richer in butter than that of the shorthorn. By a careful selection of breeding animals with a view to early maturity, the longhorns might yet regain much of their ancient popularity. We have seen the effect produced by a cross with the Longhorn bull on the common-bred dairy cows of the country. We would much like to see a cross between a good pure-bred Shorthorn bull and a Longhorn cow; such a cross could hardly fail to produce a superior animal, at least for the purpose of the graziers. At the shows of many of the leading agricultural societies, where there are no separate classes for Longhorns, they are placed at great disadvantage by being compelled to compete with other established breeds, the system of comparing the merits of totally distinct races of animals, each of which are specially cultivated to fulfil widely different purposes, is always a difficult and generally a disappointing undertaking.

CHAPTER X.

SUSSEX CATTLE.

BY A. HEASMAN, ANGMERING, ARUNDEL.

THE ORIGIN AND DATE OF THE SUSSEX CATTLE may be a matter of uncertainty. Was William of Normandy attracted by the fine oxen grazing in the rich marshes of Pevensey, or did he import them? It is generally understood they date back to the time of the Conquest, and it is well known that Pevensey and the surrounding district has always been their principal home.

This useful class of stock were formally bred principally for draught purposes, being converted into food for the public after they had cultivated the soil of the Weald of Sussex and Kent—some of the heaviest tilled land in the kingdom—and at times being required to start the heavy carriage of the county member from the same muddy district, when it was necessary for him to attend to his parliamentary duties, before locomotive power came into operation or the Highway Act had been amended. Even in those early days the Sussex cattle were fully appreciated, and, always possessing the finest quality of flesh, were never neglected by the grazier.

When they had been worked for several years, and age at last rendered it necessary that they should be drafted from the team, the farmers of the western part of the county would pay a visit to their brothers in the east; attend the fairs held at Battle, Lewes, or on the borders of Kent, in order to buy up the aged oxen; and, after grazing them a year, supply the markets with animals weighing from one hundred and eighty to two hundred stone.

Times have very much altered, and the Sussex beasts are no longer what they were, neither are they reared to the same extent or for the same purpose. They have given place to horse and steam power, and take up their position as one of the useful and established breeds of the kingdom to meet the pressing and increasing demand for beef. Their colour was formerly both light and dark red—in some instances so dark that it almost amounted to black; but the intermediate or cherry colour is now the favourite, denoting good flesh and better quality for fattening.

The breed has been too well appreciated by the tenant farmer to be allowed to die out, and great pains and attention have been taken latterly in endeavouring to alter the style and type by breeding from the smallest bone with the greatest amount of flesh; this seems to have been successful when we compare the present animals with what may be called the old-fashioned sort, one of which was fattened many years ago at Burton Park, near Petworth, and called the Burton ox. A portrait of this animal was dedicated to the gentlemen of the county of Sussex by Mr. Spilsbury, of Midhurst; its height was 16½ hands, and it measured 8ft. from the back of the horns to the tail; and from hip bone to hip bone, across the back, 2ft. 8in.; the depth of shoulder, 4ft. 7in.; girth behind the shoulder, 10ft.; and it weighed 287st. 4lb. Although this was considered a wonderful animal at the time, the meat was not in the right place; its bone was enormous, its back rib shallow instead of deep, with a spare thigh and small twist, and it was not to compare with the class of cattle now exhibited, saving in the matter of weight, which has always been a great feature in the breed. The Sussex cattle are second to none as regards early maturity and weight for age; this is proved by the weights of the animals shown at the Smithfield club meetings. The Sussex are great favourites with the butcher and consumer. At three years old well-fed steers will weigh from twelve to fourteen score pounds per quarter. Their general features may be described as follows: Nose tolerably wide; muzzle of a golden colour, thin between the nostril and eye; eye rather prominent; the forehead rather wide; neck not too long; sides straight, and not coarse at the point of the shoulder; wide and open in the breast, which should project forward; girth deep; legs not too long; chine bone straight; ribs broad; loin full of flesh and wide; hip bones not too large, but well covered; rump flat and long; tail should drop perpendicular; thigh flat outside and full in; the coat soft and silky; with a mellow touch.

The Sussex cross well with any breed, by using the male animal substance and firmness of flesh are imparted, and the colour of the offspring is generally red. They are of themselves a hardy breed, and have been found to surpass all others in the poorest pastures of their native county. The cows are not good milkers; those with the heaviest flesh are the worst, but produce sufficient to rear their calf. The most successful way of breeding is to calve them down in October and November, let them have their own calf through the winter, which can be weaned in the spring, and another calf put to the cow. If managed in this way, each cow will rear two calves, and the number of barrens be greatly diminished, which is one of the greatest evils when cows are allowed to drop their calves all the year round.

Great credit is due to Mr. Edward Cane, of Berwick Court, for the energy he has displayed in the improvement of the breed. Mr. Cane at one time was one of the largest breeders, and always ready to give a good price for the best cow brought to the hammer. He was, however, very unfortunate with his yearlings, and after losing a great many, was induced to sell off his herd. A cow purchased at this sale produced one of the best steer specimens of the breed, which was in 1867 exhibited at Smithfield, obtained the first prize in its class, and was one of the most formidable competitors for the cup, finally awarded to Mr. Combie's celebrated Black Prince. In the report of the show published in *The Field* of Dec. 14, 1867, this animal is thus alluded to: "The older class of steers contains eight entries, and is decidedly good, Messrs. Heasman's entry being probably their *chef d'œuvre*; so good that it remained out a long time competing with Mr. Foljambe's steer as second-best male." And again: "The Duke of Sutherland's very smart, level, and well-fed young Shorthorn was much fancied for second place; but he gave way to one of the best Sussex steers yet shown." Mr. Cane was the first to introduce the Sussex cattle to the notice of the Smithfield Club, and from that time much improvement has taken place. The Smithfield Club have been very liberal in their support by offering good prizes, the result of which is that the classes are well filled, and the breed year by year becomes a more prominent feature in the Christmas gatherings. The breeders will be glad to find that at the next meeting they will be put on equal terms with the other established breeds. This alteration will be of great advantage, as it will let in the autumn-bred calves. These calves are usually the best, and have hitherto been excluded.

The Royal Agricultural Society might have been expected to hold out a helping hand to an improving breed; but, contrary to this, they have taken from their list the few small prizes once offered.

The Southern Counties Association, on the other hand, are anxious to promote the breed, and give good prizes, for which they were rewarded at the Guildford show by having a large collection.

The Sussex men do not feel inclined to spoil their best animals by over-feeding.

The Public Herd Book of Sussex Stock (without which no breed is perfect) has been established about fifteen years, and promises to be of great assistance. Each year it becomes of more importance, and as recorded pedigrees increase it will enable breeders to select a cross with a degree of certainty which has hitherto been a difficulty. It will also help the sales. The public are thereby assured of the pedigree of the stock they purchase, instead of having to rely on the statement of the vendor. The book at the present time numbers one hundred and seventy-six bulls, and one thousand two hundred and eighty cows. It is also a chronicle of all pedigree prize animals, for breeding purposes, and records the names of the following breeders, who have gained honours at the Royal Agricultural, Southern Counties, or Sussex, and Kent County Shows, viz.: the Rev. J. Gould, of Burwash; Mr. H. Hughes, of Woodgate; Mr. Tilden Smith, of Beckley; Mr. Thomas Child, of Slindfold; Mr. G. Jenner, of Udimore; Mr. R. Hawes, of Westham; Mr. Thomas Bass, Messrs. J. and A. Heasman, Mr. J. Verrall, Mr. J. S. Turner, Mr. Barshall of Bolney, Mr. Wm. Botting, Mr. F. Tupper, Mr. E. Cam, Mr. G. Coote, Mr. R. H.

Burgess of Robertsbridge, Mr. J. Stoneham, Mr. J. Blencow, and Mr. A. Agate of Horsham.

The prices realised by the choicest Sussex cattle generally vary from 30 guineas to 80 guineas; the latter price was given for some sold at the late Mr. Botting's sale at Westmeston last September; while we hear of Mr. A. Agate, of Horsham, selling his five two-year-old heifers this spring at 60 guineas each, all of which are pedigree animals.

CHAPTER XI.

NORFOLK AND SUFFOLK RED POLLED CATTLE.

By THOMAS FULCHER,

ELMHAM, NORFOLK.

IN SEARCHING FOR DATA to determine the origin of this breed, we have come across a book entitled "The Rural Economy of Norfolk," by Mr. Marshall, resident upwards of two years in Norfolk (from 1780 to 1782), acting, he informs us in his preface, as agent to the Gunton Estate. Under the heading "Cattle" he says: "The native cattle of Norfolk are a small, hardy, thriving race; fatting as freely and finishing as highly at three years old as cattle in general do at four or five; they are small-boned, short-legged, round-barrelled, well-loined, thin-thighed, clean-chapped; the head in general fine, and the horns clean, middle-sized, and bent upward; the favourite colour a blood-red, with a white or mottled face. The breed of Norfolk is the Herefordshire breed in miniature, except that the chine and the quarter of the Norfolk breed are more frequently deficient. If the London butchers are judges of beef, there are no better fleshed beasts sent to London market."

Suffolk cattle, according to the earliest records on the subject, were polled, and, originally, dun in colour; later on they are described as red, red and white, and brindled.

From a very early period large numbers of polled Galloway cattle were brought into the counties of Norfolk and Suffolk. There can be little doubt that these were crossed with one or other (probably both) the native races, and that thus the present breed of Norfolk and Suffolk red-polled cattle was called into existence.

The writer is by no means disposed to accept the theory propounded by the author of the article on "Scotch Polled Cattle" (page 98), that our Norfolk polls are simply red Galloways. True enough, there is a resemblance between the heads of the two sorts, each being furnished with a thick tuft of hair, covering the forehead. In the Norfolk beast this appendage will, however, be frequently composed of a mixture of red and white hair. More rarely, a large spot of pure white makes its appearance in the face. The deep, blood-red colour of the Norfolk polls is, moreover, many shades darker than we have seen in any specimens of the Galloway breed, These two peculiarities go far to support the conclusion we have arrived at —that the old native race had a due share in the concoction of the present breed. As to by whom this cross was first resorted to, we have no precise information. Marshall, indeed, mentions that long before his time polled Suffolk, Galloway, and even West Highland bulls were used for crossing with the Norfolk home breeds; but so highly did he appreciate the good qualities of the latter, that he only refers to crossing, in order to condemn the practice.

In the absence of documentary evidence, we have it on the authority of Mr. Money Griggs* of Gately (now in his hundredth year, and for upwards of eighty years a tenant on the Elmham estate), that from his earliest recollection red polled cattle were kept in the neighbourhood of this place.

One of the oldest herds in the county is that of Mr. George, of Eaton, near Norwich. The son and successor of its founder, in reply to inquiries from the writer of this article, says: "I think about the year 1800-10, my father commenced to collect what were at that time called the blood-red polled Suffolk cows; but why *Suffolk* polled I do not know, for to the best of my recollection they came chiefly from Norfolk farmers, from Mr. Reeve, of Wighton, Mr. England, of Binham; and I well remember he had a cow picked up for him by a Mr. Walne, of Foulsham (four miles from Elmham) not long after he commenced the breed, which cow was called Foulsham, and was one of the best my father ever possessed, costing 25 guineas—at that time thought a frightful price. She bred some very good blood red calves; one, a bull, was much prized for some years. After this my father went on breeding in-and-in for many years, not being able to find bulls to his liking. The only bull I recollect his buying proved a regular brute, whom, with his offspring, he got rid of as soon as he could. After this came Mr. Birkbeck's and the Elmham blood. Mr. Etheridge, of Starston, had first and last several bull calves for himself, and the late Sir Edward Kerrison, and some went farther into Suffolk. As to Norfolks and Suffolks being the same breed, I can form no opinion, except that I know they have been a good deal mixed if only through my father's blood."

In Youatt's work on cattle, published some forty years ago, appears a portrait of a very handsome cow, bred by Mr. George. The Eaton herd still flourishes, and as might be expected from such careful breeding, exhibits great uniformity of character. Mr. Reeve and his herd have passed away, and the Binham herd no longer exists; but Mr. England writes: "My grandfather came to Binham in 1792.

* On Jan. 29, 1872, this fine old yeoman died, after a few days' illness; until within a short time before his death he was in the daily habit of walking over the farm; long after ninety he would ride to meets of the Norfolk foxhounds in his neighbourhood.

I have heard my father say the polls were much improved by Mr. Reeve and my grandfather, but whether they were red or not I cannot say; they were as alike as possible when I first remembered a herd of thirty cows here, and a beautiful red. I doubt if there are any better at the present time."

Other old-established Norfolk herds are those of Lord Sondes (Elmham), Money Griggs (Gately), the late Col. Mason (Necton), Mr. Henry Birkbeck (Stoke Holy Cross), the Messrs. Hudson (of Quarles and Blakeney), Mr. Nicholson (Gressenhall), Mr. Savory (of Rudham), and Mr. Lombe Taylor (of Starston). Of herds established within the last twenty-five years the most important are those of Sir Willoughby Jones, Bart., Mr. Colman, M.P., and Messrs. Brown (Thursford), and Hammond (of Bale), all of whom have been highly successful in the show-yard of late years. Mr. Tom Brown (of Marham), famous for his Cotswolds, is getting together a very pretty lot of cattle, beginning with heifers from Elmham, and an Eaton bull. Altogether, the red polls are now bred, with more or less care, on upwards of one hundred farms in the county.

The principal herds in Suffolk are in the hands of the Earl of Stradbroke, Col. Tomline, M.P., Sir Edward Kerrison, Bart., Mr. Arthur Crisp, Mr. M. Biddell, and Mr. R. E. Lofft, of Troston.

Mr. Badham, who at one time possessed a choice herd, and may therefore be accepted as a competent authority, is of opinion that the red-polled cattle of the two counties are now precisely the same breed.

The original Suffolks are still represented by the herd at Riddlesworth.

Amongst the good qualities that may be faily claimed for the redpolls are hardness of constitution, enabling them to thrive on scanty pasturage, and to withstand the severe winters and piercingly cold springs usually experienced in the eastern counties; their milking properties are unquestionable, and they have not that tendency to go dry which belongs to the Alderney, Ayrshire, and most other breeds having a reputation as dairy cattle. It not unfrequently happens that a cow will continue to yield a good quantity of milk from one calving to another.

The unimproved "home-bred" of twenty-five years since was open to the objection of being flat-sided, thin on the loin, light in the hind quarter—in short, a somewhat rugged-looking animal generally. By the way, the illustration accompanying this chapter must be taken to represent the old sort. What the breed of the present day is like is well described by Mr. J. K. Fowler, of Aylesbury, who, in the capacity of Judge at the last meeting of the Norfolk Agricultural Association, held at Dereham, June, 1871, was pleased to say of them: "He was struck with the remarkable usefulness and value of the cattle of this district; the cows had good useful udders, so that they were likely to be capital cows for the dairy; while the bullocks had capital chines and good backs, but they were somewhat deficient in the springing of the ribs and in the hind quarters. Amongst the lot they scarcely found an animal that was not fit for a show-yard. As a Shorthorn breeder, he wished he could put some of the good points he found upon the Norfolk polled cattle on the animals which he was breeding." It would be interesting to know the particular points Mr. Fowler wished to transfer to his Shorthorns—possibly the "good useful udders" would be one of them. Although there is believed to be not the remotest affinity between the two breeds, yet many of the female polls very nearly approach the Devon type in their sweet deer-like heads and general thorough-bred appearance.

The vast improvement which has taken place in the polled breed is probably owing, in no small degree, to the liberal prize list of the Norfolk Agricultural Society, supplemented, as it is, by the cups and gifts of private donors. So recently as ten years ago Lord Sondes was almost the only Norfolk exhibitor, Mr. Badham doing battle for the sister county. Now there is no lack of competition; of late the chief prizes have gone to tenant farmers.

The Royal Agricultural Society of England at the great Battersea show of 1862, accorded a separate class to the polls. Previously to that date, and until the last two years they have been relegated to that heterogeneous collection—the class for "other established breeds." In response to a memorial from the breeders, separate classes were granted at Oxford and Wolverhampton. On each occasion the breed was so creditably represented, that it may be hoped a similar concession will be made whenever the show is held within reasonable distance of the eastern district.

Cows and heifers for dairy purposes have been sold from the Elmham herd to buyers in the counties of Beds, Berks, Bucks, Chester, Hants, Northampton, and Sussex; whilst exportations of breeding stock have been made to Egypt, Germany, (north and south), and Austria, where, strange to say, on an estate of Prince Leichtenstein, a breed of red polled cattle has been in existence from time immemorial. Also during the last two years a few cattle have been sent to the United States, and there are orders on hand at present from the same quarter.

[Since Mr. Fulcher's admirable report was written, the first volume of the Norfolk and Suffolk Red Polled Herd Book, edited by Mr. H. F. Euren, has been published, and promises, from its careful preparation and excellent arrangement, to prove of great service to breeders. The editor has hit upon a material improvement on the Shorthorn Herd Book, viz., the addition of numbers to the tribal names, and the collection of tribes and families into groups, according to the original breeders, such groups being characterised by a letter, as well as the group's name. Thus for example, all the tribes sprung from Elmham are collected together as the Elmham group, which is prefixed by the letter A. Attached to each cow's name in the Register is the group letter and tribal number—reference to the tabulated summary of herds at once shows all particulars. This will be better understood when we give the pedigree of a bull and cow:

20 BONEY.
Calved Feb. 1872; breeder, the Right Hon. Lord Sondes, Elmham; s. The Palmer 138; d. Brettenham Strawberry [A 4], by Redjacket 163.

The letter and number enables us to find out all particulars as to the dam.

BLOOMER—A1.

Calved Feb. 1872; breeder, Mr. H. Smith; *s.* Powell 143; *d.* Dairymaid [A 1].

By reference we find that Dairymaid descends from Primrose, now 26 years old, a valuable cow of the old Elmham strain. The successful nature of the undertaking may be gathered from the fact that 246 bulls, and 670 cows and heifers are recorded. In addition to the lists, there is an excellent introduction, treating of the origin and present position of the breed.—ED.]

CHAPTER XII.

SCOTCH POLLED CATTLE.

By GILBERT MURRAY, ELVASTON, DERBY.

GALLOWAY PROPER comprises the counties of Wigton and Kirkcudbright, forming the south-western seaboard of Scotland. Geologically it rests principally on the Silurian rocks of the primary formation, attaining in some parts an elevation of nearly 2000 feet. The uplands are principally devoted to pastoral purposes, to which they are well adapted; the wild humidity of the climate and peculiar nature of the soil tend to produce a luxuriant growth of the coarser kinds of grasses and herbaceous plants, thinly interspersed with patches of heath. The average yearly rainfall is 35in.

The cattle, which derive their name from the district they inhabit, are possessed of distinctive and strongly marked characteristics. We learn from authentic sources that so early as the beginning of the sixteenth century they had attained considerable celebrity, which they have successfully maintained to the present day. The Galloway is generally classed with the mountain breeds, though we think he might more appropriately be placed amongst those of the plains. With ordinary keep, the ox at three years old will weigh from nine to ten scores per quarter. To persons but slightly acquainted with the breed they are very deceptive, as they weigh heavier in proportion to their size than any other classes of cattle. The hides weigh well, and when well fed the beasts produce internally a large quantity of loose fat. The beef is tender, and has the fat and lean well mixed up together; hence they are held in high estimation both by the butcher and the consumer.

The distinguishing characteristic of the race is the absence of horns, both in the male and female; they are mostly of a black colour, though we occasionally meet with individual animals of unalloyed purity of a red or brown colour; others again, of equally pure lineage, have white faces, and are marked with white under the belly. The trunk is symmetrical and well formed, supported by short legs;

the flesh is evenly distributed over the frame, even down to the knee and hock; the shoulders are thrown well back, which gives breadth to the chine, and depth and expansion to the chest; the sides are long, ribs well sprung; the hips round, and less prominent than in some of the improved races; the skin, though thick, is mellow to the touch, and covered with a profusion of long silky hair.

Down to the first decade of the present century, the Galloway was by far the most numerous and important breed of cattle inhabiting the wide pastoral range, stretching from the river Tweed on the east to the Irish Channel on the south-west. At the same date they were the principal breed cultivated in Carrick, the south-western division of Ayrshire. Long prior to the date of railways, and before artificial manures were known and appreciated (which gave an impetus to the cultivation of root crops), a large and important cattle trade was carried on in the south. Some idea may be formed of its magnitude at the period of which we write from the fact that, taking Dumfries as the great mart and principal centre from which the southern droves took their departure, from this one point alone there passed yearly from 25,000 to 30,000 head to be finished off on the rich pastures of Leicestershire or Northampton, or the salt marshes of the eastern counties. Several weeks were occupied on the journey; the drovers, being well acquainted with the country they traversed, in order to save turnpikes and the feet of the animals, and at the same time snatch a morsel of food, preferring the devious windings of the cross-country lanes to the more direct lines of the turnpike. There were likewise a goodly number fattened off on the rich pastures of Milton, Houwell, and Baldoon, in Wigtonshire, and on the sheltered and productive lowlands of Dumfriesshire.

One of the distinguishing features of the breed, the absence of horns, has by some been attributed to the physical conditions by which they were surrounded. The practical man who has closely studied the different races of our domesticated animals well knows that soil and climate produce wonderful effects in altering their appearance. This, and a careful selection of males and females of approved type through many generations, could not fail to stamp the race with a distinct character.

The polled breed of the eastern counties of England, which is now attaining such well-merited repute, raises a lingering suspicion in the mind, strengthened by the fact of long-continued and extensive importations of Galloway cattle into those districts, that at some distant period it derived its origin from that source. It is true that the colour is distinct, but any person wishing to improve or establish a race of a red colour would have experienced no difficulty in selecting pure-bred animals of that colour to start with. There is still a striking resemblance in some points between the two races, more particularly in the formation of the head; and it is clearly within the range of probability that the existing difference of appearance may have arisen from the effects of soil and climate influences.

Down to within the last forty years, large numbers of cattle were bred in Ireland bearing a close resemblance to the native Galloway, with the exception of horns, which the greater portion of the Irish-bred animals possessed. As many as 10,000 of these were yearly landed at Portpatrick, a small seaport on the south-western promontory of Wigtonshire, at a period antecedent to that at which the harbour was made available for the admission of loaded vessels of from sixty to one hundred tons burden. The practice was then to let go the anchor as near to the shore as the depth of water would permit without endangering the safety of the ship, and then to sling the cattle into the sea and swim them ashore. With a rough tempestuous coast this often proved a hazardous, and not unfrequently a disastrous undertaking; at all times it required the services of several well-manned small boats to guide them ashore, for if left entirely to themselves they were as likely to swim out to sea as they were to make for the land. The greater part of the cattle landed here were from one to two years old. As soon as they were sufficiently rested they were driven on to some market or fair. Nearly all fresh arrivals were pitched at Stranraer, that being the nearest market to the port of disembarkation, and at that time the great mart for Irish cattle in Scotland, and was regularly attended by graziers from long distances. In those days, and down to a much later date, the father of the writer was constantly in the habit of riding on horseback out of Ayrshire, a distance of thirty miles, to attend this market, returning home the same night. The great object of the purchaser was to select those most nearly resembling the native Galloway. As already stated, great numbers were possessed of horns; some were very small and loose to the touch, appearing to be only attached to the skin, whilst many were of a fixed and more prominent character. The matter of horns were considered of minor importance compared with that of hair and colour, and the closeness in general *contour* with which they approached the approved type. In order to obviate the difficulty and remove the objectionable appearance, the Irish breeders and cattle dealers had recourse to the following stratagem: the animal was cast in the usual way, and when properly secured, a sharp knife was passed round the base of the horn, severing the skin and fleshy part; a fine-toothed handsaw was then inserted into the incision, and the horn and pitch cleanly sawn off level with the skull. To prevent hæmorrhage, the protruding ends of the arteries were either secured by a needle and fine silk, or else they were seared with a hot iron. A circular patch of coarse linen, corresponding in size to the base of the horn, was dipped in moderately hot tar or pitch, and placed on the wound; three thicknesses of this were generally used, and stitched round the edges to the hair, to prevent them peeling off before the wound had healed up. This was undoubtedly a cruel operation, yet, when skilfully performed, we believe it was seldom attended with fatal results. It is only fair to state that the operation was only performed on those animals more nearly resembling the native Galloway. At that time many of the Irish-bred cattle had a line of white or ashy brown along the ridge of the back; nothing could be gained from removing the horns of such, as the merest tyro could at once recognise their lineage.

It has been asserted by some that the Galloway breed was greatly deteriorated from its having been crossed with the Irish. This we believe to be a fallacy, as the Galloway breeders had always a strong aversion, I may say hatred, to the Irish; the prices obtained for the pure-bred Galloways would of themselves be sufficient to deter them from adopting such a course. These cattle, when from one to two years old, would be of the average value of 4*l*. to 9*l*. per head, and always realised from 1*l*. to 2*l*. per head above the Irish of corresponding size and age. Occasionally an Irish cow might be found in the hands of a small cottager too poor to purchase a Galloway, but, as a rule, they were never used by the farmers for breeding purposes. It was then the universal custom to spay all the heifers at the age of one to two years, with the exception that occasionally one or two of the most promising were reserved to take the place of those removed by death or accident, or which increasing years had rendered unfit for the duties of maternity. There was no fixed principle of drafting out the old cows; if good breeders, they were kept to the age of ten to twelve or more years. All the Irish heifers were also spayed, either before they left Ireland or shortly after their arrival. The distinguishing mark of a spayed heifer was a triangular piece cut from the top of the right ear. The advantages gained by spaying was that the animals rested better, and consequently fed faster, and steers and heifers could with safety be grazed together in the same field.

Galloway cows have the character of being bad milkers, due in a great measure to mismanagement. The old breeders invariably allowed the calves to suck their dams, depending more on the amount realised by the stock reared than they did upon that obtained by the produce of the dairy. Farmers possessing tolerably large herds were content if they could only obtain sufficient milk to supply the wants of their family, and furnish them with butter and a small quantity of skim-milk cheese; we have frequently seen the maid milking on the one side of the cow, whilst the calf was sucking on the other. During the summer and autumn they were generally milked out of doors in the open yard. When the calf had reached the age of five or six months, it was usual to place a muzzle on the nose; this muzzle was simply a head stall, the nose piece of which was a broad band armed with sharp spikes, the motive of which was to set the cow kicking whenever he attempted to suck, and therefore defeat his object; at milking times the muzzle was removed, and the milkmaid and calf started on equal terms. When the milk became exhausted before the appetite of the calf was appeased, it became rampageous, and not unfrequently charged and overthrew the milkmaid and her pail. The cows and their calves were never separated, and except in cases where the means of restraint were used which we have related, they had free access to their dams at all seasons. Every practical man well knows the deteriorating effects such treatment would produce on the most prolific milker.

When regularly milked in the usual way we find them little inferior to many of the other established breeds, and the milk is invariably rich in butter. It has been noticed in highly cultivated cattle that the calf's sucking the dam prevents the latter becoming again in calf; whereas in animals less improved, it appears to exercise but little effect, as the Galloways as a rule were regular breeders.

In former days, cattle of all ages were generally wintered in the fields. They received a foddering of rough hay or straw twice a day. During the winter months they made little progress; as the season of grass advanced they grew and improved rapidly. Within the last thirty years the entire system has undergone transformation as if by the wand of the magician; the improvement and reclamation of large tracts of waste land, the extended growth of root crops and of the cultivated grasses, have led to the introduction of the Ayrshire breed of dairy cattle and the feeding of cross-bred sheep in large numbers, and the decline of the native race of polls.

A cross has been tried between the native Galloway and several other established breeds. That which is viewed with most favour by practical men is the cross between the pure Galloway bull and Ayrshire cow; they have the reputation of being kindly feeders, arrive early at maturity, and attain a profitable size. The prepotency of the male has recently been freely discussed; we think none of the native races are equal to the Galloway in this respect. Put a well descended Galloway bull to a herd of Ayrshire cows, and every calf will come black without an exception; but put the purest-bred Shorthorn bull which it is possible to obtain to a herd of Galloway cows, and the calves will be dropped of all imaginable colours. A good many will be black, some black and white, and a large number of a blueish-grey colour. At the present moment I have a pure-bred Galloway cow, which was put to a purely descended white Shorthorn bull, bred by Lord Zetland; the produce was a bull, which of course was castrated. He is perfectly black, and possesses all the characteristics of the Galloway. None but a person thoroughly conversant with the race would believe that he was in any way related to the Shorthorn.

POLLED ANGUS OR ABERDEENSHIRE CATTLE.

By "SCOTUS."

In the present state of farming operations much attention has of late been paid to the breeding and rearing of cattle, and a natural question has arisen, what is the best stock for the farmer to give attention to, both as a "fancy" and for marketable purposes? Different opinions on this point, no doubt, are held, but, considering all the "pros" and "cons" regarding the black polled cattle of the North of Scotland, and with such an object in view, there can be no doubt that the above-mentioned breed holds a very high place.

Though in many points they bear a strong resemblance to the Galloways, yet it has come to be the opinion of many, well versed in such matters, that they are two distinct breeds.

The Galloways, from being natives of the South of Scotland, came consequently to be better known, from their proximity to the Border and nearer access to the London markets. And, inasmuch as they were bought by English butchers and graziers, and perhaps thus known as the only black cattle in Scotland, they came naturally to represent the whole "genus."

Some years ago the experiment was made, with a view to improving the stock, to cross the two breeds together, but somehow it was a failure, as, after the first off-throwing, the stock were of a very inferior character.

As to which of the two breeds is the oldest, no one has been able to ascertain, but, as far back as one can learn, the two breeds have been separate and distinct.

Black polled cattle were first known in the counties of Aberdeen and Forfar as the Buchan and Angus "doddies," and are now not only largely bred in these two, but also in Kincardine, Banff, and Morayshire. Many points of diversity can be noticed among them in regard to size of the ears, the quality of the skin and hair, but still they are bred as one and the same.

The breeders of this black polled stock have very much increased of late years, and are still gaining ground, though the cattle plague, some years back, made sad havoc among many large and valuable herds. Among such I may mention Lord Southesk, who on that account retired from the list of breeders, though he still continues an active patron and supporter of all who are. Among other noblemen and gentlemen who give encouragement to the breeding of polled cattle are Sir George Macpherson Grant, of Ballindalloch; Mr. M'Combie, M.P., and his cousin, at Easter Skene; Mr. Morison, of Bogrice; the Earl of Airlie, Mr. Skinner, of Drummin; Mr. Walker, Portlesken, &c.

The points of a thoroughbred polled Aberdeen and Angus bull or cow can be seen in their colour being glossy black, a clean cut head, with not too great a length between the eye and nose, the former bright and prominent, and a good breadth between the two; the chest ought to be full and deep, legs short and clean boned, and supporting the body easily although firmly. The back must be straight and level, from which the ribs must spring with a gentle and easy curve. The tail must hang straight, with no protruding from behind, and finished off with a plentiful tuft of hair. Regarding the skin, it must be soft and pliant to the touch, and covered with a crop of luxuriant and silky hair. Of course, between the cow and the bull a difference must exist in the formation of the head, which in the former ought not to partake of the broad and bullet shape which that of the latter possesses, but has to be of a more elongated shape; in the shoulders, which ought to be sharper at the top; the cow also is not so broad across the chest. In both cases horns are inadmissible; and the animals being well and firmly set on their legs, should walk with an easy, springy, and "thoroughbred" looking action.

The polled breed, for their bulk, weigh heavier than Shorthorns and crosses, and command the longest prices. They are hardy and well suited from their light make to stand rough and hilly pasturage, and will thrive in a climate where Shorthorns, still less Ayrshire, could not come so fast to maturity. In the consideration, then, of such a question as this—what class of animal breeds most surely stands rough climate best, is hardiest, least liable to disease, grows and fattens soonest, and for its amount of keep weighs heaviest, and realises most to the farmer? the answer must be that, in our northern counties, these polled Aberdeen and Angus breeds come by far the nearest to such a standard. In the rearing of pure polled stock one caution should ever be present with the breeder, and that is, never to commence a herd by breeding from a purchased cow, for though she herself may seem to have all the points of thorough breeding, yet at some previous time she may have accidentally or otherwise been served by a Shorthorn or cross-bred bull; and however pure any other polled bull may be to whom she is afterwards put, the risk always is there that she may throw calves with a cross-bred strain in them. Yet, in order to keep up the strength and quality of stock, a judicious blending of different tribes and strange blood is absolutely necessary; for such, however, the greatest care must be taken to employ a bull of undoubted purity, and not to use any one about whom the slightest suspicion can exist regarding his blood or pedigree.

In order sometimes to do away with any coarseness or local defects, breeding in and in may be resorted to with success; but such a course in all breeding stock ought to be the exception, not the rule. Speaking about pedigrees, I may quote a remark, taken from a masterly paper on this same subject, not long ago read at a meeting of our north-country farmers, which is this, that farmers for their own profit ought to pay far more attention to the pedigrees of their stock than they do. For if one would consider, from many instances in England and in America, that a farmer who has his stock carefully pedigreed often gets over 100 guineas or more for such an animal which to the butcher would only value about 20*l.*, or even less.

And now let me say one word about crossing these polled cattle with other breeds. The end in view with this is simply the rearing of heifers and bullocks for the butcher *alone;* and in all cases let these be out of a black polled cow and a well-selected Shorthorn bull, combining, by such an intermixture, the neat figure and healthy disposition of the one with the size and bone of the other. Such come sooner to a marketable profit than any pure-bred animal, inasmuch as the excellencies of two superior breeds are combined together in one.

The reverse has and is being tried—a polled bull with a Shorthorn cow; but somehow from this no one seems hopeful of a satisfactory result.

Moreover, in all such crosses of different breeds no one ought to go farther than the *first* cross, as after that a decided coarseness of style appears—long legs and "weedy appearance," and none of the good feeding properties so desirable in an animal that is to be of any profit to the farmer.

I remarked before that the authorised colour of the polled Aberdeen and Angus cattle is black, yet in many cases we find them with a stain of brown down the back, and about the ears. From such some think that the colour of many of the original " Buchan doddies " was red and brindled, and, from the black colour being preferred, it has so far disappeared that this is all that remains to indicate that such had once been the case. Cows with such a strain in them are invariably excellent milkers.

In the rearing of the young, and the general treatment of this breed, the same rules hold good as with others, so that with the usual necessary amount of care, warmth, and regularity in the feeding, any one who has a fancy for what is most useful, and at the same time ornamental, will find both qualities combined in these descendants of the " Buchan doddies."

CHAPTER XIII.

THE AYRSHIRE BREED OF CATTLE.

By GILBERT MURRAY,

ELVASTON CASTLE, DERBY.

WHATEVER IMPROVEMENTS OR ALTERATIONS were made in the breed of live stock during the unsettled state of Scotland prior to the Revolution in 1688 arose either from accident or from natural causes, as the principles of breeding were then unknown. The powerful influences which soil and climate exercise upon every species of live stock, more particularly on those which are constantly exposed to the elements, are so great as to have fixed the breed of animals in every quarter of the globe. So completely is this the case that, though great improvements have been effected by the superior intelligence which nature has conferred upon man, yet through all the different varieties of live stock we trace the distinguishing peculiarities of their several districts which were originally stamped on them by Nature. The connection between the soil and climate and the cattle that are reared and fed in each situation is so intimate that they cannot be separated; at the same time we do not deny that they may be greatly altered by artificial means.

The breed under consideration is indigenous to the county from which it derives its name. The county of Ayr is divided into three separate districts; that of Carrick, the southernmost, embraces the whole of the county south of the river Doon; Kyle, the central division, occupies that part lying between the rivers Doon and Irvine; whilst Cunningham stretches north of the river Irvine to the confines of the county. This latter division claims to be the cradle of the improved Ayrshire dairy breed of cattle.

Prior to the year 1780 the inhabitants of this part of Scotland passed through a long period of religious feuds and dissensions, entailing on the people a great amount of suffering and privation. Possessing education and intelligence considerably in advance of the age, the farmers of those days were ever the warmest defenders of their homes and their religion; even at this distant date many a solitary moss-covered stone is still held in reverence, and marks the last resting-place of one who had fallen in support of the cause he had espoused.

Until the country emerged from this condition but little attention could be given to any branch of husbandry. The cattle in the best parts of Cunningham were then of small stature and badly fed, they were mostly black, with white spots on their faces, back, and other parts of their bodies; the cows had high-standing crooked horns, marked with very deep ringlets at their base, a true indication of their meagre fare. The improvement of the Ayrshire breed dates from about the year 1780, first by a cross with a stranger breed, combined with a better system of feeding. Aiton, who wrote a survey of the county, and who was himself a farmer in the district of Cunningham, could well recollect the appearance and condition of the cows in that district as far back as 1766. After great pains taken to inquire into the origin of the present celebrated breed, he was of opinion that they are descended from the native stock of the district, changed in their colour, and partly in their shape and qualities, by being crossed with the Teeswater or Dutch breeds. It is impossible to trace out all the crosses that were made between these strangers and the native cattle of Cunningham, and even to say explicitly who it was that first brought them into the district. In 1750 the Earl of Marchmont purchased from the Bishop of Durham several cows and a bull of the Teeswater or some other English breed, of a light brown colour, spotted with white; these his lordship kept for some time at his seat in Berwickshire. Bruce Campbell, who was then factor on his lordship's estates in Ayrshire, carried some of the breed into Kyle; from thence their progeny spread throughout the county. A bull from this stock was sold to Mr. John Hamilton, of Sundrum, who raised a numerous herd from that strain. About the same date Mr. John Dunlop cultivated the stronger breed at Dunlop House in the Cunningham district. However valuable the breed has now become, it is said the first offspring of the cross was far from being of the best shape. The race was chiefly propagated by coupling bulls of the stranger with cows of the native races, and, as the former were far superior in size to the latter, as might naturally be expected the progeny had at first an ill-shaped mongrel appearance, with bones large and prominent; but these cattle soon toned down, accommodating themselves to the state of the pastures; and the improvements that began about that time to be made on the soil of the western counties rendered the pastures capable of supporting much heavier stocks.

The most desirable quality of dairy cows, of any breed, is that they should yield a large quantity of milk in proportion to the quantity of food they consume, and that when dry they should feed quickly. The pure-bred Ayrshire certainly excels all other in the former, and as to the latter, she is no way inferior to many of the best established breeds inhabiting these islands. Of the quantity of milk which an average Ayrshire dairy cow yields it is difficult to speak with precision; there is not only a great diversity between some of those animals and others, but the quantity and quality of the food, the size, age, and habit of the animal, distance from or to the time of calving, all exercise a marked influence on the quantity of milk yielded at any given time. Whatever is said on the subject is open to contradiction by such as are disposed to cavil. Aiton, in his survey of the county, says that some of the dairy cows in Ayrshire yield for a time from five to six gallons of milk per day. Such returns are, however, rare; yet many, when in their best plight and well fed, will yield four gallons per day for three months, and during the season produce a total of 800 to 900 gallons per cow. Many will, however, not yield more than half that quantity, and probably 600 gallons per cow during the year may be taken as a fair average of the Ayrshire dairy stock. Aiton goes on to say that since the publication of his survey the farmers have satisfied him that he has underrated the produce of their cattle, and that they have furnished him with satisfactory proofs of various cows having produced from six to seven gallons per day for several weeks; these, he remarks, are, no doubt, extraordinary returns.

Then, as now, the farmers were in the habit of letting their cows to dairymen at a fixed rent per head, the farmer furnishing the dairy plant and the necessary food for the stock, the dairyman performing the whole of the labour. At that time the rents were from 15*l*. to 17*l*. 10*s*. per cow per annum; the calculation then was that 30 gallons of milk produced 24lb. of marketable cheese, or 12¾lb. of milk to each pound of cured cheese. Descending to our own times, the following is the result of a milking competition held at Ayr on the 26th and 27th days of April, 1861: viz.,

Name of owner.	Greatest milking.		Average of four milkings.		Weight of butter.	
	lb.	oz.	lb.	oz.	lb.	oz.
A. Wilson	28	12	24	3½	2	2
J. Hendrie	26	0	24	5	2	14¼
W. Reid	25	7	20	8¾	2	9
W. Reid	30	15	27	5½	3	6½
R. Wallace	28	14	28	8½	1	9½
R. Wallace	25	5	23	8¼	1	15

In this case, the greatest yield at a single milking was rather over three gallons, which produced at the rate of 15lb. of butter per week. We have here no record of the quantity of milk required to produce 1lb. of cheese. In the Derby cheese factory, where the milk of three hundred and sixty cows was manufactured into cheese during the year 1871, taking the average of the season, 11¼lb. of milk produced 1lb. of cured cheese.

The most approved points of the Ayrshire cow are: Head small, but rather long, and narrow at the muzzle; eye small, but quick and lively; horns small, clear, and crooked, and placed wide apart at their base; neck long and slender, tapering towards the head, with no loose skin below; shoulders thin; fore-quarters light; hind-quarters deep and large; back straight, and broad behind; the joints rather loose and open; carcass deep; pelvis capacious, and wide over the hips, with round fleshy buttocks; tail long and thin; legs small and short, with well-bent joints; udder capacious, broad, and square, stretching well forward, but neither fleshy, low hung, nor loose; the milk veins large and prominent; teats short, and all pointing outwards, and at considerable distance from each other; skin thin and loose, and the hair soft and woolly; the head, horns, and all those parts of least value should be small, and the general figure compact and well proportioned.

The Ayrshire farmers prefer their dairy bulls to possess the feminine aspect in their heads, necks, and fore-quarters, with broad hock bones and hips, and full in the flanks; they likewise pay particular attention to the formation of the small teats of the bull, and also to the colour of the scrotum. If this were any other colour than white, though the animal might otherwise be possessed of great merit, he would immediately be rejected by the best breeders.

The farmers of Ayrshire have long devoted great attention to the improvement of their dairy cows. When cows are kept solely for the dairy, and are profitable in proportion to the quantity of milk they yield, self interest would stimulate the farmer to acquire the most correct knowledge of cultivating the desirable qualities in his stock. If one cow excelled in milking, they would look out for others in which the leading characteristics were fully developed; they would rear the calves of the best milkers, knowing that they would to some extent inherit the good qualities of their dams. It has been chiefly by these means, and not by changing the stock or crossing with bulls of other breeds, that the Ayrshire dairy stock of the present day has attained its unrivalled perfection.

The improved breed was first planted in Carrick by a Mr. Fulton, in the year 1790; and in 1802 the first herd was established by Mr. Ralston in Wigtonshire, on the south side of Lockryan. They were introduced into Dumfriesshire towards the end of the last century, gaining a footing on the estate of Mr. Hope Johnstone, of Annandale. Dairy farming has spread rapidly in the south-western counties; in these parts the Ayrshire breed is gradually taking precedence of all others. *The Ayrshires bred and reared in Galloway are generally longer, thicker on the chine, rounder in the chest, and heavier in the fore quarters, and less capacious behind, than the native bred Ayrshire; they seem better fitted for the purpose of the grazier and the butcher than that of the dairymaid, thus furnishing another proof of the effects of soil and climate on the natural propensities of animals.

The superiority of the Ayrshire for dairy purposes is now generally admitted; they are to be found in every county from John o' Groats to the Land's End. The demand on Irish account is steadily increasing; hitherto the Irish farmer, as a rule, has devoted more attention to breeding and feeding than he has to the products of the dairy. For many years large numbers of the best animals have been exported to America, where they are said to succeed remarkably well; and as a proof of the value attached to them on the other side of the water, the Americans have established a herd book, in which the pedigree of all the best-bred animals is entered. A herd book in Ayrshire is much wanted; a true record of how each animal was descended would enhance its value considerably. At present the price of ordinary dairy cows ranges from 14*l.* to 21*l.*; there is always a keen competition for the best class of cows. Many dealers hold commissions to purchase all the most promising animals for exportation; show cows often realise from 50*l.* to 70*l.* each. Where so many excel, it would not only be invidious, but would occupy too much space to enumerate the breeders' names. We would strongly recommend those who wish to see the true representatives of the race to attend one of the agricultural shows which are yearly held in the county, and feel sure the most fastidious would be gratified.

There are many interesting features in dairy management peculiar to the south-western counties of Scotland. The cows are frequently let to men who either pay a fixed rent per cow, or deliver over to the farmer a stated weight of cheese; these men are provincially called "bowers." The farmer owns the cow, and furnishes a stated quantity of food, the bower and his family performing the whole of the manual labour of feeding and attending to the cows and making the cheese. On many of the Ayrshire dairy farms there is a very limited area of permanent pasture, many of the farms being under arable culture, and managed on a five or six course rotation; the cows are principally pastured on the one or two years' seed layers, which on good land keep a large quantity of stock; we have known twenty-four imperial acres of second year's seeds to pasture twenty-two Ayrshire dairy cows and a bull from the first of May to the end of September. The Scotch dairy farmers, as a rule, use hay very sparingly; on most farms oat straw is substituted, and of this they have an abundant supply. When the cows are let to a bower, the usual allowance is from five to six tons of roots per cow, in about equal proportions of swedes and common or Aberdeen turnips, and 2½cwt. of bean meal to each animal. The rent per cow varies in accordance with the quality of the pastures and the merits of the herd; from 3 cwt. to 4 cwt. of cheese per cow when rendered in kind, and from 12*l.* to 14*l.* per cow when paid in cash, are the average rates which are now obtained. On many of the high-lying farms, where the land is less suited to arable culture, with a large breadth of inferior land in permanent pasture, and at a low rent per acre—on this class of soil, with the present high prices of stock, it is now considered to be the most profitable system of management to combine dairy farming with stock rearing; hence on many farms of this description the Ayrshire cows are crossed with a polled Galloway bull, and the whole of the produce reared on the farm, and either sold off as stores to the grazier, or made off fat to the butcher, at from two to two and a half years; the crosses prove kindly feeders, and attain from eight to nine scores per quarter, at the ages mentioned above. On some of the better qualities of land, the Shorthorn bull has been used to cross the cows; with good keep the produce attain heavy weights at an early age; they are considered less hardy than the Galloway cross. There is no breed of dairy cattle in these islands that will produce an equal quantity of milk, butter, and cheese from a given quantity of food with the pure-bred Ayrshire.

CHAPTER XIV.

WEST HIGHLAND CATTLE.

By JOHN ROBERTSON, Blair Athol, N.B.

AT THE PRESENT TIME the breeding and rearing of cattle has become a specially important subject. The increase of the population, the wonderful prosperity of trade, and the consequent high rate of wages, have combined to increase the consumption of animal food to a very high degree, while the supply has not at all increased in proportion; and, with the present high rate of wages, in agriculture as well as in all other departments of industry, but with no corresponding rise in the price of grain, farmers will be driven to abandon tillage to a great extent, and to take to the production of animal food, in which branch of their calling they have now the only hope of coping successfully with the foreign producer. It may not, therefore, be unprofitable to glance at some of the sources whence our supplies of animal food are derived. For the last twenty or twenty-five years the breeding and rearing of cattle has, from various causes —chief among which are the abolition of the corn laws and the development of the railway system—become a promiment feature in Scotch farming, and perhaps the best proof of the success with which this department of agriculture has been cultivated, not only as regards the native breeds, but breeds more properly English, is the immense traffic in fat cattle which has sprung up from Scotland to the London Market. The breeds of feeding cattle chiefly reared in Scotland are the Shorthorn, the Polled, the West Highland, and crosses.

Of the pure Scotch breeds, that known as the West Highland, though not now so numerous as it once was, nor so highly prized as some of the southern breeds, deserves special attention, not only on account of its being the original breed of the west, and a great part of the north of Scotland, but also on account of its importance in point of numbers, its quality as food, and its value as a breed to cross with. It is very difficult to trace the origin of any distinct breed of cattle, because climate, soil, mode of treatment, and other conditions combine, through time, in impressing special physical characteristics; but there can be no doubt that the horned, shaggy, hardy, and comparatively small breed of cattle, now best known as the West Highland, has for ages been the breed peculiar to the mountainous district of Scotland, although it is now chiefly confined to the counties of Argyll, Inverness, Perth, and Dumbarton, if "confined" is a word which can properly be applied to so extensive a territory and so untamed a breed. A well-bred animal of almost any species, is a pleasing object; but there are perhaps fewer animals familiarly known to us so graceful in form, colour, and movement as a thoroughly well-bred Highland ox or heifer. In form it possesses all the characteristics so much and so justly prized in the Shorthorn—the straight back, the short legs, the broad chest, the breadth of loin and depth of rib, and, in short, the "squareness" and solidity of form, which always imply weight, whether in man or beast; while the noble branching horns, the fine, full, and fearless eye, the short, broad, well-bred muzzle, the shaggy coat of richest black, or red, or dun, or brindled colour, impart a picturesqueness which is still further enhanced by that grace and deliberation of movement so distinctive of all animals reared in perfect freedom. All these characteristics of the breed are frequently found in the Highland oxen exhibited at our Christmas shows; but there the most attractive appearance does not carry the prize. The more sentimental and less earthy points, however much they may denote purity of breed, are overlooked by matter-of-fact judges of fat stock, and the prize goes—very properly perhaps—to the fattest, but not to the finest beast.

Until about eighty or a hundred years ago, the mountains of the north of Scotland were pastured with "black cattle," as the West Highland breed is still frequently termed, when they were gradually displaced by flocks of sheep. Previous to that time the stock of a Highland farmer consisted of cattle, horses, goats, and a few sheep, the goats and sheep being generally penned at night; and the remains of those pens are now frequently seen in the form of grassy mounds or the foundations of ruined walls on the green sites of the old "shealings," or summer quarters of the Highland pastoral farmer of days now long gone by. The cattle, from their hardy character, lived on the higher mountains during summer and autumn, and during winter and spring subsisted as best they might on such rough herbage as they could find among the woods or on the meadows of the lower straths or valleys; but a frequent result of this rude manner of farming was that a severe winter, or still worse a severe spring, cut off by sheer starvation a large proportion of the stock. The only consolation was, that at that time the stock did not represent much money, and the rent was not difficult to pay. It is not above a hundred years since a grazier in the district of Rannoch, in Perthshire, not reckoned at the time as by any means a wealthy man, lost from starvation, one severe spring, one hundred and twenty head of cattle. Now, the system of management, if system the old mode of farming could be called, is entirely changed. In many parts of the Highlands sheep have entirely displaced cattle—only a few cows, sometimes only one, being kept on the farm for domestic purposes, and these few are frequently Ayrshires or crosses. But over the greater part of the mainland and islands of the counties of Argyll and Inverness, the northwest of Perthshire, and the highlands of Dumbarton and Sterling shires, West Highland cattle are bred and reared on the lower lands, generally with marked improvement and success, and in many instances to great perfection. And there is every reason to believe that, in the districts named, this breed of cattle is the most profitable to cultivate, because from its hardy character it will thrive, both in summer and winter, under circumstances in which the smoother coated and softer breeds would perish. Highland cattle are easily fed, and the quality of the beef is admittedly superior, and consequently in great demand; but the objection to them, as compared with softer breeds, is that they do not come to maturity so early, nnd consequently do not yield the same quick return of capital, and the objection is so far perfectly sound. Therefore a lowland farmer finds it more profitable to breed and rear Shorthorns or polled cattle or crosses than Highland cattle; but a Highland farmer rears his West Highlanders at little or no expense beyond the value of the hill or meadow pasturage in summer, a great deal of which is valueless otherwise, and of such meadow hay or straw and turnips as they may get in winter, until they are fit to send to market at six quarters or two years old; and then he can afford to sell them at such a price as will enable the southern farmer to buy them and carry them on to profit until they are fit for the knife. They are not so well adapted, however, for court or yard feeding as they are for the open field or for tying up, and that is a disadvantage.

In former times a very extensive trade was carried on, chiefly by "drovers," who bought Highland cattle from farmers and at district fairs in the north, and sent them to England and to the southern counties of Scotland—the great mart for this trade being the Falkirk Trysts, which are still held, as of old, in the months of August, September, and October, and at which the southern dealers and farmers meet their brethren from the north. Within the memory of men not yet very old, before railways or even fast coaches were in existence, and before the peculiarities of districts and races were so much effaced as they are now rapidly becoming, the scene presented by a Falkirk Tryst was a very animated and striking one. There, year by year, met crowds of men remarkably different in form and face and dress and language—any one of whom might be selected as a good type of the shrewd and active man of his peculiar race or district—to transact business on a large scale in the open field, amid the strange din of lowing cattle, barking dogs, flourishing of sticks (and sometimes even more), frantic torrents of purest Gaelic, broken English, vigorous Lowland Scotch and English Saxon; the whole scene not unlike a mimic Flodden. But times have changed all this, the improved means of communication having in a great measure rendered "trysts" or meeting places unnecessary; and neither for cattle nor for sheep is Falkirk Tryst any longer the national institution it was in bygone days. A story is told of a remarkably shrewd Highland grazier of the last century, who from his wealth and position in his own country was called "The Baron," and who had a fine fold of "black cattle." The Baron told a friend that he was going to Falkirk Tryst with cattle. "What!" said his friend, "*you* go to Falkirk to sell cattle, without a word of English in your head!" "Never mind," said the Baron; "I have no English, it is true, but my oxen will speak for me."

The English trade in Highland cattle has, from various causes, such as diminished supply and preference for other breeds, very much declined; but great numbers are still bought for the rich pastures of England, on which they fatten to perfection, and Highland beef so fed is deemed the finest that an English gentleman can place on his table.

The largest folds of Highland cattle are in the islands of Uist and Skye; but in all the islands of the west coast of Scotland this is the breed almost exclusively reared, and in no other part of the country are its leading characteristics so fully developed. The nature of the pasturage, the moist climate, and the comparatively mild winters consequent on vicinity to the sea, produce hair and horn such as the more inland pastures of Perthshire can never rival; but, on the other hand, the inland pastures are supposed to conduce more to the growth of bone than do the island and seaboard pastures, and consequently the cattle reared in the inland districts are generally much heavier. Perhaps one of the finest herds ever seen in Scotland was in the island of Harris, thirty or forty years ago, in the possession of Messrs. Donald and Archibald Stewart, who, by judicious mixing of the best blood that could be got in the counties of Inverness, Argyll, and Perth in their day, cultivated the breed to notable perfection; and their three-year-old oxen of large size, with horns like buffaloes and hair like goats, used to attract great attention in the districts through which they were usually driven to Falkirk Tryst. This herd is now chiefly represented by that of Mr. John Stewart, at Duntulm, in Skye, whose cattle always carry high honours at the national shows. Mr. Macdonald, Balranald, in Uist, owns perhaps the largest fold of pure Highlanders in Scotland, there being above a hundred breeding cows with their followers. This fold is of long standing and of great note, both as to numbers and quality. They are not heavy cattle, but fresh blood from the best Perthshire herds has been of late introduced, which will no doubt improve the stock in weight. Captain Macdonald, of Waternish, in Skye, and Dr. M'Gillivray, in Barra, have large and well-managed herds of fine cattle. Lord Colonsay had, in Colonsay, a well-bred herd of long standing, which has lately been dispersed under a change of management. Thirty or forty years ago this herd was wonderfully improved by a bull, bought at what was then thought a very high price (120 guineas), out of a Rannoch fold in Perthshire. In Mull, Jura, and especially in Islay, there are fine herds of Highland cattle, the rich pastures of the island last mentioned being very fattening. On the mainland of Argyll no Highland fold is more noted than that of Poltalloch, and for many years Mr. Malcolm of Poltalloch's name has always been found among the first-prize takers at the great shows. This stock has been carefully and judiciously managed and improved by periodical importations of fresh blood from the best inland herds, chiefly from the Breadalbane stock. Indeed, throughout the higher grounds of Argyll the class of cattle which one sees on almost every farm is very superior, and there is no better place for getting good Highland heifers than the June fair at Dumbarton, which is the great market for Argyllshire Highlanders. Doune Tryst, in November, is perhaps the best market for Highland "stots" or oxen. On the mainland of Inverness-shire Highland cattle are not so much bred as they are in Argyll, probably because the climate and pasture generally are less adapted for cattle than sheep, and, although the pastures on many farms would rear very fine cattle, they are generally reserved for winter feeding for sheep. In the valley of the Spey, however, from Cluny Castle to Grantown, the cattle are chiefly Highland, and on many farms well bred, but of late years Shorthorns are creeping into Badenoch and Strathspey among other southern fashions and innovations, and for crossing purposes it is probably a judicious change; but, without doubt, black cattle are better suited than any other to the meadows of the Spey and the climate of Badenoch. The Earl of Seafield has in recent years established at Castle Grant a fold of Highlanders which is rapidly coming to fame. A year or two ago he secured a remarkably fine bull from the Duke of Athole's herd, which had carried the first prize in the class of aged bulls at the Edinburgh Show in 1869. Mr. Fraser, at Faillie, near Inverness, owns a herd which has long been noted in the north, and which is maintained in high efficiency by periodical purchasers from the best folds in the country. The Faillie "stots" are noted, and usually bring the highest prices going in the north for such cattle. In the county of Perth Highland cattle have diminished very much in numbers within the last forty or fifty years, but there is now a reactive tendency, and they are again receiving attention in the upper glens of Athole, Breadalbane, and Balquhidder. Before the period just named there were numerous remarkably fine herds in the districts of Callander, Balquhidder, Breadalbane, Glenlyon, and Rannoch, and the names of Messrs. M'Laren, Callander; Macdonald, Monachyle; John Stewart, Donald Stewart, and Charles Stewart, Glenlyon; M'Laren, Rannoch, &c., were familiar over the Highlands as famed breeders of stock. During the gradual dispersion of these celebrated herds there was selected from them with great care and judgment the nucleus of that famed herd owned by the late Marquis of Breadalbane, which, at the time of his death in 1862, was probably the finest in Scotland. For many years Lord Breadalbane took a personal interest in his Highland cattle, and both in their selection and management he had the assistance of his friend and neighbour, the late Mr. Stewart Menzies, of Chesthill, than whom there was not perhaps in Scotland a better judge of Highland cattle. The Breadalbane stock was carefully drafted every year, and the annual October sales afforded, for many years, an excellent opportunity to farmers and other breeders of improving their stocks by purchases of pure blood. When the Breadalbane herd was sold in 1863 the principal purchaser was the late Duke of Athole, who then founded, or rather engrafted on an old stock—for there were in Blair Castle the remains of a traditional breed of white Highland cattle—a herd which has since well maintained at Blair Athole the fame acquired by the Taymouth fold. Mr. Malcolm, of Poltalloch, was also a buyer at the Breadalbane sale; and the folds of the Duke of Athole, Mr. Malcolm of Poltalloch, and the late Mr. Peter of Aberfeldy, probably represent among them all the excellent points of the Breadalbane stock. The sale of this stock attracted much attention at the time, and the prices there given were very unusual for Highland cattle. The late Duke of Athole bought the celebrated red bull Donald at 135*l.*, which was some years afterwards sold to the Hon. Lady Menzies, Rannoch Lodge. The late Duke of Hamilton bought a fine three-year-old dun heifer at 126*l.*, which was afterwards sold to the late Duke of Athole, and which, as the property of the present Duke of Athole, stood the first prize cow at the Highland Society's Show at Inverness in 1865. These prices, however, are trifling in comparison with the prices realised at the sale of the Dunmore Shorthorns in 1872, which should encourage the Highland breeder to persevere, and to hope that his favourite breed will some day become proportionably, if not equally, fashionable and valuable. The great feature in the Breadalbane herd was weight combined with fineness of quality, and the Athole herd still maintains that character in a marked degree. Many of the animals exhibited from this fold at the Highland Society's shows of late years had, with perfect purity of blood, all the characteristics of the Shorthorn as to shape and size. A Highland three-year-old ox of the Athole stock, exhibited by the Duchess Dowager of Athole at the Perth show in 1871, was regarded by eminent judges at the show as the perfect model of an ox in shape.

The management of Highland cattle varies considerably in different districts, and according to the size of the fold. In the larger and more reputed folds the cattle are at large summer and winter, the breeding cows only being placed in separate pens or sheds at calving time, which is usually from January to March or April. In winter the stock generally get meadow hay or straw, and in many cases a few turnips in the open field, in addition to what rough grass they pick up in woods and other sheltered places; and it is surprising how they maintain condition under such treatment during the most severe winters. For some time after calving, and until the young grass comes on in May, the calves are kept separate from their dams, and let in to them to suckle three times a day; but when the cows are turned afield the calves are turned out along with them, and remain at foot until they are weaned, which is usually done about the beginning of October. The experiment of allowing the cows to calve in the open field, and letting the calf follow the dam at will from birth has been tried; but the result was that both cows and calves became very wild, and the cows very dangerous to approach. In general Highland cattle are gentle and good-tempered; but when left to roam at large in the woods or on the hills, where they seldom see the face of man, or at least of a stranger, they become shy, and, like all wild animals, guard their young with jealous care; and the means of offence and defence at the command of a Highland heifer are not to be lightly regarded by the most courageous. In some good folds, in Inverness-shire for example, the breeding cows are housed and milked like dairy cows, and the calves reared by hand; but this is done only on mixed farms, partly arable and partly pastoral. A Highland cow yields nothing like the quantity of milk that an Ayrshire does, but the quality is much richer. The age at which Highland cows calve is usually four years, because it is found that, unlike softer breeds, the heifers are not at maturity until they are three years old, and of course breeding at an earlier age stops their growth. The usual practice with Highland farmers is to draft off in October or November their old cows and surplus young stock, the latter generally at six quarters old. Prices of course vary with demand and quality; but from 8*l*. to 12*l*. is the ordinary range of prices for the better sort of this class of young cattle.

The crossing of Highland heifers with Shorthorns is a subject which is often discussed, and generally viewed with great favour by good judges of both breeds of cattle, but the experiment does not seem to have been yet tried with such success as to have commanded much attention. There may be various reasons for this, but it occurs to us that a main cause is that the experiment has hitherto been chiefly, if not exclusively, tried by southern breeders, crossing two-year-old heifers or aged cows with shorthorn bulls, producing in either case a diminutive offspring. If three-year-old heifers were brought direct from the hills and crossed with a pure-bred Shorthorn, and afterwards maintained on their usual "sober" fare, there is every reason to expect that the result would be satisfactory; and no cross is so likely to be useful in upland districts as this, combining, as it should do, the "growthy" qualities of the Shorthorn with the hardiness of the Highlander.

CHAPTER XV.

THE GLAMORGAN BREED OF CATTLE.

By MORGAN EVANS.

SOME OF OUR OLD BREEDS OF CATTLE are rapidly disappearing, and it is well to note their existence ere it is too late, and to record their merits and failings whilst any trace of them remains. The exertions of a few eminent breeders have risen some of our indigenous cattle into world-wide repute. Other once celebrated breeds have not been so fortunate, and have been all but crushed in rivalry with contemporary animals that have been cultivated with more care and further developed towards perfection in shape, size, and quality. The tendency of modern agriculture is to obliterate local breeds of farm stock. Improved farmyards, improved systems of cropping and manuring land, gradually lead to the adoption throughout the kingdom of improved breeds of cattle to the exclusion of purely local strains. The Shorthorns, Herefords, and Devons are ever extending the boundary of their influence, and the counties from which they originated are not now, as at one time, the sole homes of these cattle. Where high farming is practised, one of our fashionable breeds is generally adopted. Great hardihood may be dispensed with when cattle are never exposed to cold and rain. Early maturity and rapidity in fattening when in the stalls are the qualities most sought for. Wherever a country is in a highly advanced state of cultivation, the hardy native oxen of the district become abandoned for breeds more suited to the commercial interests of the farmer in a time when quick returns on his capital are of vital importance. Rough waste pastures give way to broad acres of wheat and turnips; active grazing cattle are consequently replaced by large docile animals, that accumulate great weight of flesh when kindly treated in stalls or boxes. The plough is no longer drawn by oxen, but by stalwart horses, or by the more powerful aid of steam; and the modern farmer, instead of selling all his cattle from the summer grass, to be turned into profit by the grazier, manufactures meat himself by the aid of cake and corn, and sells the produce direct to the butcher.

Among the breeds doomed to extinction is the once so well known and highly prized Glamorgan breed. The Glamorgans are of ancient lineage, and their origin is hidden in the past. They belong to the class called middle-horns, and in character and antiquity of descent they rank with the Herefords, Devons, the Welsh black cattle, and other allied breeds. As far back as the twelfth century, it is said that a Norman knight, Robert Fitzhammond, who had seized a great portion of Glamorgan, introduced some Normandy cattle into the county, which are supposed to have been crossed with the native cattle. The swelling crest of the Glamorgan ox is by some traced to the influence of this admixture of foreign blood. Youatt says that the influence of Devon blood could not be mistaken at the end of the last century, and attributes it to the importation of Devons into the district by Sir Richard de Grenaville, one of the knights who at one time divided the lordship of Neath. It is certain, however, from all legends and historical accounts, that the Glamorgan cattle are a very old breed, and were the native cattle of the district from a very early date, and that their principal characteristics remain unchanged, Norman and other breeds notwithstanding. In more modern times the Glamorgan farmers were particularly careful of their breed, and we are told that in the last century they prided themselves greatly on the fact that they admitted no admixture of foreign blood into their cattle. The Glamorgan cattle soon became famous. Stock in England at that time were fed at grass. There was no stall feeding and no improved Shorthorn. There was land to plough, and active strong oxen did most of the work. The ideal of a good breed consisted in the females being hardy and profitable milkers, and the males active, docile, and strong workers in plough and cart, and beasts that, when their allotted period of farm labour was done, would, at six or seven years old, fatten into brave oxen on the broad English pastures, on their way up to London and other great centres of the beef-eating population. The Glamorgan breed was celebrated for these desired qualities, and about the commencement of the present century they were highly prized and much sought for by the great English graziers and feeders in the counties of Northampton, Warwick, Wilts, and Leicester. George III., who has been dignified by Youatt with the character of being a "good judge of cattle," was very partial to this breed. He stocked his farm at Windsor with them, and periodically recruited the herd with fresh blood from the Welsh country fairs.

Notwithstanding the high patronage of a king, and other circumstances which might be thought favourable to their development, they have gradually declined in character and in numbers, until at the present time there is no pure herd of these cattle to be found in the county where they were so long held supreme. The Glamorgans are almost extinct. A cow here and there of the old type might be found, but they have greatly degenerated in size and quality; and a pure-bred bull of the true sort it would be difficult, if not impossible, to find. The reason of the decline of these once famous cattle is popularly attributed to the high price of corn during the French Revolution and the succeeding wars of Napoleon, which were followed by the breaking up of the old fine pastures of Glamorgan for the purpose of growing greater breadths of grain crops. Why the Glamorgans should succumb under such influences more than many other well-known breeds I cannot say. It is, however, certain that from that time less care was bestowed on them, and they diminished in number. Consequently their fame became more circumscribed, and when the farmers of the county once more turned to breeding cattle, they took advantage of the fashionable improved breeds that had already gone so far in advance of their native stock.

The Shorthorns, Herefords, and Devons had stolen a march on the Glamorgans. The native breed still held its own for a long time in the dairy, being much superior to either of its rivals in milking qualities. A few energetic breeders now rose to do battle for the cattle of their forefathers, and although very great progress was made, and a fair standard of perfection at certain points was attained, it became impossible to stem the tide against the invaders. Three or four local breeders were prominent to the last, but death and other changes caused the last strongholds to give way, and we might almost say the end has come.

The breeders of greatest note in late times were Mr. David, of Radyr, and Messrs. Edward and Christopher Bradley, of Treguff. Respecting what was commonly known as the "Treguff breed," Mr. Edward Bradley—who at the age of eighty-six enjoyed lively and pleasurable reminiscences of days gone by when he championed his favourite breed—kindly recounted its history to me in a letter which lies before me. When a boy he had often heard conversations by practical men on the superiority of the Glamorgans, and a regret expressed that they had not been taken more care of for the purpose of the shambles, instead of being yoked to the plough in the farmer's team; and he found himself some sixty years ago in a position, with the aid of a younger brother, to enter into an undertaking towards the restoration of the breed of Glamorgan cattle. "The origin of the Treguff breed" (he says) "was purchased from a mountainous district in this county, and she possessed great valuable points to be admired as desirable towards establishing a herd likely to become valuable. This cow was selected as having a frame to be desired for the dairy as well as for breeding. She was comparatively a small animal, showing a capacious udder and a perfectly formed body, straight over the back and loins, deep chested, wide hips, short legs, and a particularly small bone. Her hips and shoulder joints were round—such as are seen generally in the Hereford, Shorthorn, &c.—showing the greatest aptitude to fatten quickly. Her colour was an admixture of brown, a bay or red. This cow was in calf, and was the founder of the Treguff breed, which in a few years established its celebrity at the provincial shows. My brother, who knew well the characteristics of the different points requisite towards furthering the value of the breed, took advantage of every opportunity to purchase cows having qualifications likely to improve the stock. The Glamorgans for several years became great favourites. They were valuable in the dairy, and no meat in the butcher's stalls showed more quality, nor would any other breed vie with them in thick marbling throughout the carcase. Many of them were fattened at three years old for the butcher, and the Treguff cattle have taken upwards of sixty prizes at the Tredegar and other shows against Herefords, Durhams, Devons, &c. A prize given at one of the Tredegar shows for 'the best three-year-old beast of any breed' was awarded to my three-year-old, weighing nineteen score a quarter, and he was sold to a Newport butcher for 53*l*. After my brother's death, and at the expiration of the lease of Treguff farm in 1850, the stock were all disposed of by auction; the breeding portion became crossed with such description of animals as were at the time in the purchaser's possession."

The Glamorgan cattle produced a rare quality of meat, highly prized in the metropolitan and provincial markets. They were profitable to the butchers, being well lined inside with tallow, and their meat, from its first-rate quality, always commanded the highest price. The average weight of cows, says one who wrote in 1814, was at that time from eight to ten scores, and oxen from twelve to fourteen scores per quarter. Youatt thus describes the breed: "They were of a dark brown, with white bellies, and a streak of white along the back from the shoulder to the tail. They had clean heads, tapering from the neck and shoulders; long white horns, turning upwards; and a lively countenance. Their dewlaps were small, the hair short, and the coat silky. If there was any fault, it was that the rump, or setting on of the tail, was too high above the level of the back to accord with modern notions of true symmetry."

Martin says they were a superior breed, "generally of a red or brown-red colour, often with white faces, and otherwise varied with white. The head was small, the aspect lively, the neck inclined to be arched, the carcase round and well turned, the back rising to the root of the tail, which was peculiarly elevated." That the Glamorgans had white faces is incorrect. Youatt also errs in a similar way, for he illustrates his description of them with a drawing of a white-faced cow from the Royal farm at Windsor. No pure-bred Glamorgan ever had a white face. The most fashionable colour for a Glamorgan cow was an admixture of a rich brown with red. The bulls were invariably black, with, of course, the usual white markings, and many of the cows were of the same colour.

The Glamorgan breed at one time extended through the counties of Monmouth and Gloucester. There can be little doubt that the old Gloucester cattle were descended from the Welsh stock, and they may therefore claim a notice here. They also, like their progenitors, are almost extinct. The only existing herd of pure Gloucesters is that kept by the Duke of Beaufort at Badminton, where they were first established nearly a century ago. Thirty years ago there were two other herds in Gloucestershire, one that of Mr. Leonard Stanley, and the other that of the late Col. Kingscote. The former was sold off in 1843, and the Kingscote stock in 1852. The Badminton herd now alone remains. Several cows and heifers were bought at the sales of the Stanley and Kingscote herds to introduce fresh blood into the Badminton strain; since that time no change of blood from the old sort has taken place, but about fifteen years ago four heifers of the best Glamorgans that could be obtained were purchased, and bulls bred from them. The cross did not effect any change in their character or colour, but reduced the size materially. The true Gloucester cow shows a good deal of character, being a lengthy, good-looking animal, light fore, but deep hind quarters, with good milking points. The body is brown; head, nose, and legs black; well-shaped white horns with black tips; tail and top of rump (or ridge of tail) white; white udder with black teats; the upper side and end of tongue is also black.

The great peculiarity is the white mark, which extends from the loin along the ridge of the tail, and down between the hind legs to the fore part of the udder. Only a sufficient number of calves are reared at Badminton to keep up the herd, which numbers about fifty head; consequently the draft cows are sold to the butcher. I am indebted to Mr. John Thompson—who has so ably managed the Badminton estate since 1842—for much of the above information. He adds, in writing to me, "Although good, hardy animals, they fatten slowly, which I think is the chief cause of the breed not having extended; but when fat the meat is very good, having a larger portion of lean meat to fat than most animals, and are considered by the butchers 'good cutters;' but I have no doubt a well-bred Shorthorn would make 3lb. of meat from the same food that a Gloucester would require to make 2lb."

Several attempts have been made to improve the Glamorgans by crossing with Ayrshires, Herefords, and Devons, but without success. A breed so old, and of such a fixed type, was not likely to blend harmoniously with any other. Improvement could only come from within; all attempts to improve by crossing must and did end in failure.

CHAPTER XVI.

PEMBROKESHIRE OR CASTLEMARTIN CATTLE.

By MORGAN EVANS.

IF WE LOOK for the oldest breed of domesticated cattle in Great Britain, we must evidently turn to the western coast. With the descendants of the ancient British people will be found the descendants of the ancient British cattle. The successive invasions on the east and south coasts of our country of conquering Roman, Saxon, Dane, and Norman, drove the aboriginal inhabitants with their cattle into the west and far north of the island. There can be no doubt that the Welsh and the Scotch Highlands are the oldest of all our existing breeds of cattle; not even excepting the white cattle in Chillingham Park. The oldest colour I believe to be black, notwithstanding much popular tradition to the contrary. From the mention of white cattle in certain ancient records, it has been too readily assumed that the oldest native stock of the country was white. The laws of Hywell Dda, or Howell the Good, written in the tenth century, certainly speak of white cattle; and the Dimetian code says that "the privilege of the Lord of Dinevwr is to have for his *saraad* as many white cattle with red ears as shall extend in close succession from Argoel to the palace of Dinevwr, with a bull of the same colour along with each score of them." Youatt says that the "same records that describe the white cattle with red ears speak also of the dark or black-coloured breed." I do not know the passages in the records to which he alludes. There is a strange absence of any mention of the colour of cattle in the old Welsh chronicles, and I cannot find a single passage to the purpose in the writings of the bards, where one would naturally look for some information.

The distinct specification of colour in the Dimetian code supposes the presence of cattle other than white. It may even be assumed, as far as the above proves to the contrary, that white cattle were not the common stock of the country at the time, although they appear to be the sort most highly prized. The great favour in which they were held might even be because of their rarity; possibly also because of real or traditional value attached to them; or, as has been conjectured, they may have been larger than the blacks. The tribute, however, appeared to demand a special rather than a common kind of beast. There need have been no selection of a particular colour, unless cattle of another kind coexisted with them, and, indeed, prevailed to a greater extent. There is no doubt of the great antiquity of a race of wild white forest beasts. They are alluded to very distinctly, in 1598, by John Leslie, Bishop of Ross, and also by Hector Boece, in his "History and Chronicles of Scotland," quoted by Professor David Low in his work on the "Domestic Animals of the British Islands." And, according to Speed, King John received from Maud de Breos 400 cows and a bull, all white and with red ears, as a present to his Queen, in order to appease his Majesty, whom her husband had offended. This variety, hitherto considered the oldest, is of very early origin, yet still possibly an offshoot developed from the older blacks, and perpetuated in certain centres by natural selection. Their present descendants, the white cattle of the parks, frequently throw black calves, appearing to revert to the earlier type. Mr. Darwin evidently doubts the great antiquity generally attributed to them, for the following passage occurs in his work on "The Variation of Animals and Plants in Domestication," p. 85: "The cattle in all the parks are white; but, from the occasional appearance of dark-coloured calves, it is extremely doubtful whether the aboriginal *Bos primigenius* was white;" and he also adds that certain facts which he enumerates show "that there is a strong, though not invariable, tendency in wild or escaped cattle, under widely different conditions of life, to become white with coloured ears." There is a general belief in Wales, as in England, that the old breed of the country was white, and many specimens of this sort might have been seen very recently in the Principality. Indeed, Low, who published his superbly illustrated book in 1842, selects as a specimen of the wild or white forest breed a drawing of a "cow eight years old from Haverfordwest, in the county of Pembroke." Cattle of this kind are now very rare in Pembrokeshire. The breed of the country is black, and known sometimes as the Castlemartin, but now more generally as the Pembrokeshire breed. There can be little question of the great antiquity of this breed. "The Pembroke race in England," says Mr. Darwin, "closely resemble in essential structure *B. primigenius*, and, no doubt, are its descendants."

Youatt says, "Great Britain does not afford a more useful animal than the Pembroke cow or ox." There is no breed which for general usefulness can successfully compete with the Pembrokeshire cattle in their native district, they are so perfectly adapted to the climatic and physical characteristics of the country, and to the system of farming generally practised there. The country is hilly, and some of the best pastures are greatly exposed to the storms of wind and rain so frequent in the autumn and winter months. Farmyards with sufficient accommodation for breeding and feeding, after the English fashion, are extremely rare. The landlords are somewhat to blame in this matter for not providing the farmer with proper accommodation for man or beast. As a set-off to severe criticism of the landlords, it may be urged that the system of agriculture practised in South-western Wales is generally such as does not demand very extensive out-buildings. To the entreaties of a tenant who asked for cattle sheds, such as were built on other farms on the estate, one landlord is reported to have replied, "I'll build you a new barn and cart house, John; and when I see you grow turnips I will build you a cattle shed." This answer appeared very pertinent to the individual holder, but I think a very self-evident retort might also have been made; at least, no better inducement to growing green crops and stall-feeding cattle can be given than by building the appropriate number of sheds, which the farmer always readily turns to proper use. Improved farming can only follow, and never precede, improved farm buildings, in Wales as everywhere else.

Many reasons might be adduced to show that the breeding and rearing of stock must long remain the principal feature in the agriculture of the district, and that under existing conditions the native cattle are those most profitable to the majority of the farmers. The counties of Pembroke, Cardigan, and Carmarthen are well adapted for breeding cattle, sheep, and horses. The humidity of the climate is favourable to the growth of grass, whilst the soil is firm and dry. Foot rot among sheep is almost unknown, the hoofs of the colts are well formed and hard, and very different from the spongy, flat-footed animals bred in the Fens on heavy clay soils. The bed of outlying stock is firm, and never becomes sloppy in the wettest weather, the undulating country allowing any excess of rainfall to run off freely to the swollen rivulets. The grass is not rank and coarse, but short and sweet. It is not a corn-growing country. The spring season is late, and the autumn weather commences early. Crops that are grown profitably elsewhere are liable to great damage in consequence of the lateness of the harvest and the wet weather which frequently sets in at that time. In consequence of the humidity of the climate, it is more suited to growing oats than wheat. Beans and peas are a most uncertain crop. Many attempts at growing large quantities of both have been made; the result is that beans have been entirely abandoned, whilst a few acres of peas may still occasionally be seen. But the uncertainty of harvesting this crop in proper condition will be understood when it is proverbially said in Pembrokeshire that no man grows peas more than five years in succession. High farming as in England can never become the rule here, whole parishes may be found in which there is no more than a single field that could possibly be tilled by steam. Mr. Mechi's advice for treating thin-skinned land—viz., to plough deeper—would result in imbedding tons of ploughshares fast in the eternal rocks, to be philosophised over by confused antiquarians in the next century, or the plough would turn up the "yellow rab" in such quantities as would ruin any farmer less opulent than a City alderman. In fact, all attempts at high farming—growing crops and harvesting them after the fashion in the midland counties or in the Lothians—have been delusive. All Englishmen and Scotchmen who have migrated to this part of Wales have either signally failed or succeeded only by adopting the peculiar farm practices of the country. The late Mr. Rees, bailiff to Lord Cawdor at Stacpoole Court, having been asked at an agricultural dinner at Carmarthen to give his opinion on Scotch farming, said he knew little about it, as there was "no Scotch farmer's grave in Pembrokeshire—none of them ever remained long enough to be buried there." And I remember some years ago being asked whether I ever knew a Scotch farmer who was able to hold his own in Pembrokeshire for more than seven years. I was unable to reply in the affirmative.

The above remarks will, I trust, be found a digression more apparent than real. I wish to show that, notwithstanding the undoubted pre-eminence in many respects of some of our more widely-spread and improved breeds of cattle, there is still some ground for supposing that in this peculiar locality the ancestral blacks are cultivated for sound practical reasons, and that their improvement is an object worthy of attention. Taking into account the climate, soil, and average homestead accommodation in the country, the Pembrokeshire cattle can be bred and fed cheaper than Shorthorns or Herefords. Surely an ungenial climate must tend to increase the expense of keeping a beast. Wintering cattle is dearer than letting them run the fields in summer. The more cultivated and delicate breed are under the disadvantage in Pembrokeshire of having to be housed a fortnight or three weeks earlier than the blacks, and they must be kept in later for about the same period in the spring. This makes a material difference in the estimate of cost for the year, where there is a mixed system of dairying, breeding, and feeding carried on. There can be little doubt that, in the district under notice, a herd of black cows can be kept fifteen per cent. cheaper than an equal number of Shorthorns, and still yield as much butter or cheese—two articles that form an important item in the rent-producing power of the Welsh farmer. Capitalists holding sheltered and luxuriant pastures, having extensive farm buildings, and who aim at producing large, prime fat beasts, may there, as elsewhere, keep Shorthorns to greater advantage than any other breeds; but persons of limited means living on poor land and with small farmyards, cannot do better, I think, than retain and cultivate the indigenous breed of the country. I hold there is no middle course: either blacks or Shorthorns. The Shorthorns are undoubtedly the grandest cattle in existence for early maturity, size, aptitude to fatten, and do well with proper care in any part of Wales, as in almost every part of the globe. The arguments in favour of introducing any other improved breed into the district must be futile, when even the suitability of the Shorthorn, with all its signal merits, is a question open to dispute.

The character of the Pembrokeshire breed may be better shown by comparison or contrast than by repeating the hackneyed phrases usually adopted to describe the ideal features of all oxen—whether Shorthorns, Herefords, Devons, or Welsh. The main requisites in every animal producing flesh for consumption are depth and breadth. A mention of peculiar features in virtues and failings is preferable to vague generalisation. The colour of the Pembrokeshire breed is black. The horns are of great length, white tipped with black, wide-spreading and curving upwards. The head is of medium length, longer than the West Highlands, and somewhat longer than the Devons, approaching the Herefords or the improved Sussex in form. The nose is small, and the neck fine, with little tendency to the "throatiness" observable in some breeds. The eyes are prominent, but without the untameable gleam of the West Highland or Chillingham cattle, domestication having removed any special traits of wildness and of ferocity. The coat long, not straight like the Highland cattle, but wavy, or sometimes curly. The forehead is broad. The tail is of good length. These may be said to be some of the chief characteristics of the Pembrokeshire breed in contra-distinction to other well-known cattle, although it does not very correctly represent the type aimed at by the breeders generally. For instance, in Wales no more than elsewhere is a white horn considered the best, but a yellow, mellow, and oily-looking horn, having the unction mark of a predisposition to fatten—a horn in which the black extends more than a few inches below the tips, or one that has a hard blue colour throughout—is to be condemned. Several writers have remarked on the colour of the skin as being of an orange yellow, and the coat on the barest parts of the body as being of a brownish hue. Some of the best breeders in Pembrokeshire are careful to maintain this characteristic in their herds. This, along with a yellow horn and a wavy coat, almost invariably indicates a beast that will feed well either at grass or in the stall. A short, crisp, coal-black coat is not to be compared with one that is long and wavy. The outer covering of hair put on in the winter months should, with outlying cattle, at the end of spring and during the early summer months be of a russet brown. One frequently sees cattle of this breed whose coats are one mass of ringlets; but experience, I think, shows that they are not the most easily fattened, and I do not know to what source to attribute this peculiarity. The hair on the forehead of bulls is often very much curled, and it is rather to be admired than otherwise for the sake of its picturesqueness, as well as that it indicates other important qualities.

The meat produced by these cattle is excellent, and not to be surpassed in texture and quality. The milking properties of the cows are certainly equal, if not superior to those of most modern improved breeds. I have the authority of eminent London dairymen for stating that Welsh black cows are on the average equal to any class of cows in milk-producing capabilities. The only objection to them at dairy farms around the metropolis is their colour. The admixture of black with red, and white, roan in the herd is not thought fashionable or pleasant to the eye.

There is a tradition in Pembrokeshire that the Castlemartins were improved by the importation of Devon bulls, but I do not know on what basis such tradition rests. I do not place much reliance on it, although the red colour of the Devons easily merges into black when crossed with them. Mr. Thomas Lewis, of Norchard, Pembroke, some years ago, assured me that in all the best strains of Castlemartins there was a cross of the Herefords. A Mr. Hitchings, of Ermytage, he said, had migrated to the north of Pembrokeshire to the home farm of the late Mr. Barham, of Trecwn, and brought back with him to Ermytage, in Castlemartin parish, on his father's death, a lot of half-bred Herefords, and black cows in calf from Mr. Barham's bull. The best breeders, according to Mr. Lewis, adopted the infusion of strange blood, and he says that the most celebrated stock in Castlemartin were soon after impregnated with it.

I have seen and closely watched several attempts to improve the breed by crossing. An effort to introduce Hereford blood ended in utter failure. The white face was difficult to wipe out, and the progeny of the cross never appeared to work into the blacks. The Devons amalgamated more easily, and the produce were more of the proper type, but did not on the whole improve its general character. The West Highland cross were small, useful little beasts, and, though they sold very readily at the fairs, were not as profitable as the native breed of the district, for there was a loss in size and in milking qualities, with no corresponding advantage gained. Several Pembrokeshire farmers have tried a cross with the Anglesea cattle, with no good effect—although I must confess that Mr. Richard Harvey, of Haverfordwest, has produced remarkably fine stock by this cross; but I believe the same spirit of enterprise which he has shown in trying to improve the breed by going to North Wales might have been equally, if not more, successful, had he turned his attention nearer home.

The quickest and best way to improve the breed is by making a judicious selection of the old stock of the country. By breeding from the best and weeding the bad, considerable progress might be made in developing it to a perfection hitherto unattained. It is not impossible that a superior class of animal might be produced by judicious crossing; but the path is one beset with difficulties when we have to deal with a breed so distinct and of such a unique character as the Pembrokes. Holding this view, I, in 1867, proposed the formation of a herd book as the only safeguard and guide to improved breeding. The opportunity given was but lightly esteemed, for no more than three or four breeders put forward any claim to making an entry in the book. I also proposed the formation, on joint-stock and co-operative principles, of a herd of no less than twenty of the best cows that could be obtained (with, of course, the requisite number of bulls) to found a herd farm, where scientific principles in breeding were to be adopted, and where an annual sale of the surplus stock was to be made. The estimation in which joint-stock companies were held at that time was fatal to the project; but I still think it to be the most important and feasible ever made to develope the Pembrokeshire cattle. It must be understood that no single breeder was to be found who would attempt the tactics of a Bakewell or a Collings. I therefore proposed a combination of effort, which, had it met with proper support, would in one decade have raised the Pembrokeshire breed to a position they are not likely to reach for a quarter of a century.

As good useful graziers, the Pembrokeshire cattle are justly celebrated. It is to be hoped that they will not too soon pass away and become extinct. Pembrokeshire farmers should do all in their power to improve them. The formation of a herd book is a necessity. I am glad to find the project revived by Mr. J. B. Bowen, of Llwyngwair, late M.P. for the county, who deserves success for the praiseworthy tenacity with which he clings to the improvement of the native blacks.

A few errors current in popular descriptions of the Pembrokeshire cattle have to be corrected. For instance, Youatt says that a "few have white faces, or a little white about the tail or udders," and that the "Pembrokeshire cow is usually black, with occasionally a dark brown, or less frequently a white face, or a white line along the back." Mr. W. C. L. Martin, in his work on "The Ox," following Youatt, commits similar blunders. No white is admissible, except, perhaps, on the udder; any other markings of white obviously denote strange blood. Even a white udder is not to be admired, and is exceptional. A coat of a brownish colour is not uncommon; indeed, a brown tinge many breeders consider an indication of aptitude to fatten, and as denoting rapid growth in their young stock. I lean very decidedly to this opinion. It is necessary to state, however, that this brown is of a peculiar hue, and the slightest tendency to red must be emphatically condemned.

The Pembrokeshire black cattle are the principal stock of the counties of Pembroke, Cardigan, and Carmarthen. The county of Pembroke is divided into two sections, the English and the Welsh speaking population occupying separate districts. The hundred of Castlemartin, the south-western part of Pembrokeshire, is along with the hundred of Roose, entirely English. The black cattle of Castlemartin were fifty years ago pre-eminent in the county. From their geographical position, and from the greater enterprise of Castlemartin farmers, the Castlemartin cattle had become famous before the other stock of the country. The farmers of North Pembrokeshire, living in the hundreds of Dewsland and Kemes, until recently always resorted to Castlemartin for their bulls, and advertisements of sales of black stock invariably held out as an inducement to purchasers that certain bulls and cows were of the pure Castlemartin breed. At the present time these cattle are characterised as of the "Pembrokeshire breed" in all such documents, and also in the catalogues of local shows. Careful breeding commenced with the farmers of Castlemartin, and the stock of the country has been much influenced by the use of Castlemartin bulls. The cattle in the hundred of Dewsland are larger than those of Castlemartin, and at the present time are of equal if not of greater merit than the strain so long considered representative of the Pembrokeshire cattle, and that gave them the distinctive name of Castlemartins.

CHAPTER XVII.

THE ANGLESEA CATTLE.

By MORGAN EVANS.

ISLANDS ALWAYS HAD a peculiar fascination to aggressive people bent on conquest; islands also generally breed heroic men, and possess a distinct fauna. The Isle of Anglesea is famous as the scene of frequent invasions, and as the home of an ancient breed of cattle. Soon after the Roman general Suetonius Paulinus had supreme authority in Britain, A.D. 58, he pushed his forces on to subjugate this little island spot of but 173,000 acres, or about 270 square miles. His troops, it is said, swam across the straits to Mona, as the black cattle imported therefrom periodically swam hitherward during nearly eighteen centuries after. The island was again attacked in the same century under the direction of Julius Agricola, who was sent by the Emperor Vespasian to command the forces in Britain in 78. Besides the numerous affrays between Welsh princes, who appear to have had a faculty for fighting with their kith and kin, and amongst other incursions of its enemies, the Danes in 900, and again in 969, landed there and made great havoc; and in 913 and in 966 the Irish, with their peculiar instinct of practising home rule by entering into quarrels with their neighbours, laid the place waste with great cruelty. In 1096 it fell a prey to English troops under the Earl of Chester and the Earl of Shrewsbury. Henry III. invaded Wales in 1245, and made a tool of his judiciary in Ireland to attack Anglesea, but, not coming quickly to the support of his minion, the Irish forces were assailed and driven back to their ships. Edward I., also, in 1277 reduced the island with a fleet from the Cinque Ports. After the inhabitants submitted to English laws and government, the natives became patriotic—perhaps to a fault; for in 1648 they had a council of war, and issued a general declaration in favour of King Charles. With the blind pluck inherent in islanders, they thought their little strike for the king would change the aspect of affairs and avert his impending certain fate. But the last humiliation of the inhabitants was at hand, when on the 2nd of October in the same year they capitulated to the Parliamentary forces under General Mytton, and, like peaceable, good-natured people in novels, they have lived happy ever after.

The island of Iona, one of the western islands of Scotland, was remarkable in ancient times for its want of cows; but the island of Mona, in Wales, has long prided itself on having a good stock of these animals. St. Columbus prohibited cows from grazing on the slopes of Iona, for he said, "Where there is a cow there will be a woman, and where there is a woman there will be mischief"; but saints and sinners agree that Anglesea was the special home of herds of prolific cows, whose progeny were transported in great numbers into the adjacent land to become food for the people of North Wales, and in later times to penetrate into the large pastures in English counties. The ancient British called their island Mon, a name the Romans Latinised into Mona. The Welsh characterised it as "Mon Mam Gymru," or Mon the mother of Wales—supposed by some to refer to its general productiveness, by others to its being possibly at one time the principal seat of learning and Druidical lore. I imagine the term "Mother of Wales" arose from its maternal capabilities in supplying food to the mainland by its corn and hordes of cattle. An old proverb says that "As Mona could supply corn for all the inhabitants of Wales, so could the Eryri mountains afford sufficient pasture for all its herds if gathered together." The Rev. Robert Ellis, of Carnarvon, celebrated as a Welsh scholar and literary antiquarian, suggests, in answer to some queries put to him by me, a derivation for the ancient name of the isle as pertinent as it is original. The word "môn" signifies cow; thus Pontfôn becomes Cowbridge, "Henfonfa" a place to keep cows; the Isle of Môn is therefore the isle of cows. However correct this derivation may be, Anglesea has certainly ever been conspicuous as a cattle-breeding country.

The productiveness of Anglesea in cattle has always been great for such a limited area. Roberts's Map of Commerce, published in 1649, gave 3000 as the number of cattle annually exported and swum across the Straits of Menai. The losses by this mode of transit were few; cattle are good swimmers. This fashion of swimming beasts has been known elsewhere along our coasts in equally difficult places. The Rev. Walter Davies, in his "General View of the Agriculture and Economy of North Wales," says that Mr. Lewis Morris, in his "Book of Charts," in 1747, puts the number of animals exported in his time as 15,000, although this must be an evidently exaggerated account. Mr. Davis, in 1810, gives the average export as "not above 8000, from one to four years old." Youatt, writing after the erection of Menai Bridge, considers he does not exaggerate when he estimates the annual export at 10,000 head, of an aggregate value of 50,000*l*.

The black cattle of Anglesea are nearly allied in character and race to those of South Wales. Mr. Darwin definitely pronounces the Pembrokes as descendants of *Bos primigenius*, but thinks, with Professor Richard Owen, that the blacks of North Wales have their origin in *Bos longifrons*. I am not prepared to discuss the osteological distinctions on which he bases his inferences. History and the geographical situation of black cattle on the western and northern coast of Britain seem to favour the idea of one common origin, and that they are the oldest breeds in the country. They are found at intervals in a line from St. David's Head in Pembrokeshire to as far north as Iceland.

The Anglesea cattle are very like the Pembrokes. The coat, as with the Castlemartins, should be long and wavy. This generally denotes good quality, and a growing beast easily fattened. In colour they are generally darker than those of South Wales, being a pure black. A little more white is allowed than in the Pembrokes, the scrotum of the bulls and the udders of the cows being very frequently white. A white streak is sometimes found along the chine, but this feature cannot be commended. The horns, which may be broadly described as white with black tips, curving gracefully upwards in cows and oxen, are usually much darker-coloured than in the Pembrokes, and the white portion not so mellow and creamy in appearance. They are perhaps a little larger than the Castlemartins—standing on short strong legs; but are not so good in the head or shoulder. The head of the ox is very frequently heavy and bull-like. Davies, in his time, attributed the "bull-like features in the head and dewlap" of the Anglesea ox to the fact that the calves were not weaned in Anglesea until "double the time at which they are weaned in other counties," together with their not being "gelt until they be about a year old;" but this will hardly account for the persistency of this feature in stock not thus treated. The shoulder is often coarse, and falls in behind the bladebone. In short, comparing them once more with their rivals, they are altogether coarser in the fore part than the Pembrokes, but have better hind quarters—wider, with bigger thighs and broader loins.

As breeding and rearing cattle has from time immemorial been the pride of Anglesea, the development of good dairy qualities in the cows has for ages been neglected. In the beginning of the present century the island but barely produced enough butter and cheese for home consumption. It is not therefore to be expected that the Anglesea cattle, under such a course of treatment and selection, should have inherited great milking powers, and consequently they are in this respect surpassed by many other breeds.

Mr. T. Congreve, at the Christmas Meeting of the London Farmers' Club, speaking as a grazier, deeply lamented that year by year it became more difficult to buy a lot of good beasts to graze. The high price of meat at the present time and the high cost of labour seem to foreshadow a time when more attention will be paid to breeding cattle. The demand for good grazing beasts is increasing, and the prices given for them at the present time remunerative. Our large dairies have relinquished breeding; our stall feeders are many of them entirely dependent on the fair and market for their stock. I apprehend that the production of good grazing young beasts and those ready for the feeder's stalls will for some years prove to be a profitable pursuit for the farmers in our breeding districts, whatever the breed of cattle they raise, provided they be of the requisite quality. Few districts are so suitable as Anglesea for stock raising; few cattle so highly prized for the quality of their meat, or thrive so well on good English pastures. They are hardy, and may be reared at little expense. For some time to come—at least as long as the present aspect of things remains—the farmers of North and South Wales who breed the native cattle will not have cause to regret any attention they may bestow on improving the native stock. And although their farmyards are generally ill-constructed, and deficient in requisite accommodation for feeding purposes, the farmers may become somewhat more reconciled to their position with the increased demand for their cattle consequent on the relinquishment of breeding elsewhere, and the scarcity of healthy animals throughout the country.

Black stock in Wales are always readily bought up by the drovers who frequent the fairs in the districts where black cattle are a speciality. They can be disposed of, half fat, at a pound or two a head more than coloured beasts of the same weight. Whilst this continues, it is some inducement to the local breeders to keep to the type and colour of their beasts—especially for those who have not the proper farm buildings for stall-feeding all their oxen and selling them only as prime fat. During last summer the demand for Welsh runts in the fairs of the Principality has been great; so much so that one breeder, writing to me after visiting a local fair, said, "the drovers were all mad."

The Anglesea cattle are now cultivated to equal perfection in Carnarvonshire and some parts of the adjoining counties as in the "mother" isle; and diminutives of this breed are the principal stock of the mountainous districts of Carnarvon and Merioneth. A few breeders have paid considerable attention to improving the Angleseas; they offer good sound material for development. Bakewell, it is said, thought highly of them in this respect. The Rev. Walter Davies naïvely says that this eminent breeder thought that " in some points they were nearer his idea of perfection in shape than any other he ever saw, *his own* improved breed excepted." But they cannot be improved by crossing with English breeds. They will not blend with foreign blood; the colour becomes destroyed and the type broken, and the produce cannot be reduced to an uniform standard. Endeavours in this direction have been fruitless. An improved type of Angleseas, if it is to be obtained, must be evolved from themselves, as must also be the case with the Pembrokeshire breed. But, as the only possible cross of the latter not evidently retrogressive is with the Angleseas, so the cattle of North Wales allow of no fresh blood except perhaps that of their kindred in South Wales; and a writer in Morton's "Cyclopædia of Agriculture" suggests this cross as the best, and one calculated to improve them in many important points.

The old white cattle are sadly becoming extinct in Wales. It may not be uninteresting to record a late attempt to form a herd of these white cattle. An English gentleman who took a farm in Pembrokeshire (Mr. Tebbitt, of Castlecenlas), collected a few white cows of the old breed common in the country at one time, and, after obtaining a white bull, continued for several years breeding cattle of this character. The experiment was not long persisted in, for he afterwards resorted to Shorthorn bulls, and crossed his stock. But some of the white cattle he produced, especially many of the cows, were remarkably fine in size and shape, and a few years ago the herd was disposed of by public auction.

In one instance only have I witnessed the appearance of albinos in a herd of black Welsh cattle. They were a dun fawn, with light eyes of a pinkish hue; some of them were very short-sighted—almost blind. These animals cropped up in a herd of the purest blood in the country. They could all be traced to be descendants of a certain black cow, likewise of the purest strain. The albinos appeared in the herd with great frequency. They were principally females, but the albino cows gave birth to calves of the usual black colour. As their offspring was always black, it did not greatly interfere with the course of breeding pure stock on the farm. It was, however, very annoying to find the tribe that had the best blood in the herd throwing albinos in every direction.

Cattle like the Angleseas and the Pembrokes should require no apology for their existence, at least for some time to come. The peculiar character of the districts in which they are bred in climate and soil, and the system of farming generally pursued in these parts, are sufficient excuse for their presence on English ground. But whenever the agriculture of Wales becomes thoroughly suited to a more improved breed it will be folly to persist in rearing black cattle because of their antiquity, or because of any other supposed virtues. All local breeds must ultimately be obliterated, except in places where they have a special calling, so to speak. When such cattle as the incomparable Shorthorns are to hand, it is the greatest bigotry and short-sightedness not to avail ourselves of them whenever it is practicable.

CHAPTER XVIII.

THE KERRY BREED OF CATTLE.

By R. O. PRINGLE,

AUTHOR OF A "REVIEW OF IRISH AGRICULTURE."

ROM THE EARLIEST PERIOD of which we have any record, the rearing of cattle has been a principal feature in the industrial pursuits of the people of Ireland. This is evident from the frequent mention made of cattle in the early annals of the country; and it is recorded that among the tributes paid to the King of Cashel in the fifth century were about nine thousand head of cattle. At a still earlier period the "inhabitants not only fed upon the flesh of oxen, but were clothed in their skins, formed weapons (pins and fasteners) out of their bones, used their sinews and intestines for strings, and employed different parts of these animals in ministering to clothing and decorative arts."

Such are the statements made by Sir William R. Wilde, in a most interesting paper "On the Ancient and Modern Races of Oxen in Ireland," which he read a few years ago at a meeting of the Royal Irish Academy. Sir William informs us that "cattle formed not only in early times the chief wealth and produce of the country, but were also employed as a means of barter. Thus we read of ransoms being paid with oxen, and as many as 140 milch cows being given for a manuscript." The names of many places in Ireland are derived from circumstances connected with the rearing of cattle, and most of these names have come down to the present day unaltered, or with a very slight variation from the original.

Sir William's paper was founded upon an examination of the heads and other remains of ancient oxen found in bogs in Ireland, and these relics, he remarks, "are exceedingly curious in an historical point of view, as they afford undeniable evidence that, so far back as the eighth or tenth century at the latest, we had in Ireland a breed of cattle which, for beauty of head and shortness of horn, might vie with some of the best modern improved races so much admired by stockmasters, and which are now being re-introduced from England." The heads of ancient oxen show that four breeds or races of cattle existed in Ireland in early times. These, according to Sir William R. Wilde, were "the straight-horned, the curved or middle horned, the short-horned and the hornless or Maol," or Moyle.

Of the latter, living representatives may still occasionally be met with in remote parts of the country; but the Kerry breed, which belongs to "the curved or middle horned" race, must be considered the sole modern representative of the ancient breeds of Irish cattle.

The Kerry cow is a handsome animal, and small in point of size. The late Earl of Eglinton, when Lord Lieutenant of Ireland, described the Kerry breed as "the thoroughbreds of cattle." The points of the true Kerry are described as follows in the "Review of Irish Agriculture," published in the last part of the *Journal of the Royal Agricultural Society of England.* "The former," that is, the true Kerry, "is a light, neat, active animal, with fine and rather long limbs, narrow rump, fine small head, lively projecting eye, full of fire and animation, with a fine white cocked horn tipped with black, and in colour either black or red." The woodcut of the Kerry cow given in Youatt on Cattle, while correct in most points, gives the idea that the breed is of a roan colour, which is never the case.

We have used the term "true Kerry" on account of certain crosses or subvarieties which exist of the breed. The first of these is the "Dexter," which, it is said, was introduced about fifty years ago by a Mr. Dexter, who was land agent on Lord Hawarden's estates. It is doubtful whether the "Dexter" variety is the result of a cross with another breed or of selection. If it arose from a cross, it is very difficult to say what breed was used, and we are rather inclined to believe that the "Dexter" variety originated in a selection of males and females having certain distinguishing characteristics, which were perpetuated in after generations by attention to the same principle in the selection of animals retained for breeding. Certainly the points of the "Dexter" variety are sufficiently marked to enable anyone to pick them out after a little experience. "The 'Dexter,'" as stated in the "Review" above quoted, "has a round, plump body, square behind; legs short and thick, with the hoofs inclined to turn in; the head is heavy, and wanting in that fineness and life which the head of the true Kerry possesses; and the horns of the 'Dexter' are inclined to be long and straight." It must not be supposed, however, that the "Dexter" is to be considered a spurious Kerry, although not an original Kerry, which is the meaning we wish to convey by the term "true Kerry." This fact is recognised at all the Irish royal shows, and, what is perhaps of more importance with reference to this point, at the shows of the County of Kerry Agricultural Society, the only difference being that there is no separate classification at the royal shows, whereas the prize sheet of the county society has separate sections for the original Kerry and the "Dexter." The latter is, indeed, a great favourite with many persons, and we have frequently observed that the majority of the prizes at shows where there was no separate classification were awarded to "Dexters."

A cross of the West Highland Kyloe blood has been introduced into the Kerry breed. This, however, has not been done recently, so far as we can learn, nor to any great extent. The traces of the cross are sufficiently evident to those who are familiar with the particular characteristics of the two breeds. The generality of people, however, either do not detect the slight traces of the Kyloe cross which exist in some Kerries, or do not attach any importance to them one way or other.

The greatest drawback the Kerry breed has had to encounter is neglect on the part of breeders, and the result is that it is often difficult to pick ten really nice animals for breeding out of a lot of fifty or sixty heifers as exposed for sale in any of the fairs in Kerry. The remainder would be good grazing beasts, but not such as a fancier would select as breeders. Still, very nice specimens are often picked up among such lots, and we have reason to believe that prizes have frequently been awarded to heifers picked up at fairs, and of the breeding of which nothing was known. In fact, entries in royal show catalogues frequently run as follows: " Kerry heifer, the property of Mr. A. B.; pedigree and breeder unknown." These animals had been bought by good judges simply because they had bred back to that character which the old breed possessed. This tendency to breed back to the old type is a feature in the Kerry which proves it to be an original or native breed of cattle, and should be taken advantage of in all attempts to improve the breed.

The Kerry cow is very docile, and is always a great pet when kept by the owners of suburban villas. She has also the property of being able to bear confinement; for instance, we have known a Kerry cow kept for five years in a stable in Dublin; she had only two calves during that time, and was scarcely ever dry, keeping up a full supply of milk for a large family. We have stated that the breed is small in size, heifers, and sometimes even cows, not exceeding 40in. in height at the shoulder. This was the height of Alderman Purdon's very neat first prize heifer, exhibited at the late show of the Royal Dublin Society. She had a calf just before the show, although only two years old, and her milk vessels were as large as those of many cows three times her weight. The dimensions of a fat Kerry cow, which was awarded a prize at a show of the Royal Dublin Society, are given in the article on Irish Agriculture already referred to—namely, " 38in. in height at the shoulder, 70in. in girth, and 42in. in length from the top of the shoulder to the tail head," indicating " a weight of about thirty imperial stones." Of course many Kerries run larger, or rather stand higher.

The Kerry cow is very easily kept, and this characteristic, combined with her milk-producing qualities, well entitles her to the appellation of " the poor man's cow," which has been bestowed upon her. The yield of milk and butter depends of course very much on the keep; but we happen to know that the average yield of milk produced by the Kerry cows belonging to a gentleman who has for many years paid great attention to the breed is about twelve quarts, or three gallons daily, and the average yield of butter from 6lb. to 7lb. per week. Some of his cows have produced more, but the quantities stated have been about the average.

These cows are no doubt well kept; but it is a large yield, considering the size of the animals and the comparatively small amount of food which they consume.

The Kerry breed is, however, not only suited for supplying the dairy, but cattle of this breed also fatten rapidly on even middling pasture, and their beef is exceedingly fine and well flavoured. This is a feature of all the varieties of the Kerry, even of the most neglected specimens of the breed, and hence it is that cattle which would not be selected for breeding are quickly bought up by graziers when driven to other parts of the country.

Considering the merits of the Kerry breed as a really useful as well as a fancy breed, it is gratifying to know that breeders in Kerry have of late been paying more attention to it than they did. The County of Kerry Agricultural Society is rendering valuable assistance in this work of improvement, and we confess we look to that Society for doing more good in this way than either the Royal Agricultural Society of Ireland or the Royal Dublin Society. Of the gentlemen residing in Kerry who have evinced most interest in the improvement of their native breed of cattle we may name Richard Mahony, Esq., Dromore Castle, Kenmare; James Butler, Esq., Waterville, which is a post town; and the Knight of Kerry, Glanleam, Valencia. Several gentlemen in other parts of Ireland have also small herds of Kerry cattle, to the breeding of which they have given careful attention. Among these are Earl Fitzwilliam, Coollattin Park, co. Wicklow; C. Brinsley Marley, Esq., Belvedere House, Mullingar, Westmeath; Capt. Bayley, Friarstown, Tallaght, co. Dublin; and Mr. James Brady, Raheny, co. Dublin, who has for a long period been a careful breeder and successful exhibitor of Kerry cattle. The blood which Mr. Brady has chiefly bred from has been that of the herd belonging to the Knight of Kerry. It was amusing to see Mr. Marley's first prize three-year-old Kerry bull, Rory of the Hills, bred by Mr. Brady, turned into the ring at the last spring show of the Royal Dublin Society, to compete with Shorthorn, Hereford, and Devon bulls for the Chaloner Plate, valued at 150 guineas, as "the best bull of any breed over two years and under six years of age." For the first time the judges declined to award the cup to the Shorthorn bull, and gave it to the Devon; but when it was observed that the judges were going to set the Shorthorn aside, many thought, and very reasonably, that the Kerry bull would have carried off the greatest honour of the show.

Sir Robert Peel and the Rev. J. C. Macdona have of late years introduced the Kerry into England; the latter gentleman, after carrying off the prize at the Royal Agricultural Show of 1871, sold his herd to Mr. J. H. Murchison, who now possesses the original of our engraving.

The prices of Kerry cattle vary considerably. Ordinary cattle, suitable chiefly for grazing, may be bought at moderate prices; but if a heifer is likely to prove a fancy animal, her value is increased a hundredfold. Prize cows and heifers frequently bring as much as 15 to 25 guineas, and we have known even larger sums refused for choice specimens. The following may be taken as ordinary rates: Common Kerry heifers, well selected, from 6*l*. to 7*l*. each; common cows, well selected, 10*l.* to 11*l.* 10*s.* Mr. James Bogue, Passage West, co. Cork, who has been long resident in Kerry, advertises that he buys Kerry cattle on commission; but, as we have already said, it is not easy to pick up superior animals, unless one travels a good deal through the county, and takes time to look about him. We may mention, however, that the principal fairs in Kerry where Kerry cattle are to be met with are those held at Killarney, Killorglin, Castlemaine, Cahirciveen, Sneem, and Dingle.

CHAPTER XIX.

THE ALDERNEY BREED OF CATTLE.

By AN AMATEUR BREEDER.

THE CHANNEL ISLAND BREED OF CATTLE, popularly known in this country as "Alderneys," consists of two classes of the same breed. The Guernsey is the larger of the two, usually of a light fawn colour, patched with white. The Jersey class is smaller, and the colour to which more attention has been paid is a dark or, as the Scottish say, "dun" deer, and is popular in England, no doubt in consequence of its more aristocratic appearance. The Alderney is essentially a cream-and-butter-producing breed, giving more milk and of richer quality in proportion to its size than any other cow; the best have been known to give from 10lb. to 14lb. per week. This merit gives them their place in live stock, either for dairies near fashionable towns like Brighton, or as cows for the park and the villa paddock, combining in the highest degree utility and ornament. The dairies of great cities are chiefly supplied by cows of the Dutch or the Shorthorn cross, which give large quantities of comparatively poor milk, and when dry fatten easily for the butcher. This is not the place of the Alderney, which, in England at any rate, is essentially the gentleman's cow.

Writers on the subject, copying one another, assume that, because the Channel Islands were once a dependency of Normandy, the Alderneys are an offshoot of the Normandy breed; but few breeds could have less resemblance. It has also been suggested that they are an offshoot of another good dairy tribe, the Ayrshires; but Ayrshires are much more like a small Shorthorn cultivated for milking purposes. At the great International Exhibition of live stock in Paris in 1855, where nearly all the ox tribes of Europe were represented, the late Fisher Hobbs, of Boxted Lodge, Essex, a very good judge, came to the conclusion that the true ancestors of the Ayrshires were Danish, and that the Alderneys were more probably descended from some Swiss mountain breeds, of which many specimens were there exhibited—dark and light fawn in colour, and fine in head and horns.

At the present time there is no doubt that in England, where the principles of selection have so long been successfully applied to horn stock and sheep, finer specimens of the Alderney have been produced than in their native islands.

For many years the farmers of the Channel Islands, while sternly prohibiting any importation of bulls, have made the rearing of heifers for the English market a profitable part of their business; but it is only within a comparatively recent period that they have learned from English breeders the advantages to be derived from a careful selection in obtaining symmetry as well as milk.

Amongst English breeders who have shown what could be done towards obtaining the best points of a milking cow by applying Bakewellian principles of selection, Mr. Philip Dauncey, of Horwood, near Winslow, Bucks, occupies, or rather occupied, the most distinguished position. For nearly half a century he devoted his attention to obtaining great milking qualities, symmetry, constitution, and a uniform fawn colour without white. His success placed him at least half a century in advance of the Channel Islanders. When in 1867 Mr. Dauncey retired from stock farming in consequence of his advanced age, his sixty-nine cows and heifers produced 3285*l*. Mr. Majoribanks gave over one hundred pounds for his cow "Landscape," and Mr. Walter Gilbey just under that sum for the heifer "Ban."

Mr. Dauncey produced a breed much more hardy than the original Channel Islanders; his stock lying out on the pastures throughout the year. The imported Alderneys are delicate, and, on first introduction, require slight shelter in the cold weather, but they soon afterwards become acclimatised.

A decided improvement has taken place in Alderneys since 1833. The Jersey Agricultural Society was founded in that year under the presidency of General Thornton, the Lieutenant-Governor. The council of the Society drew up a scale of points from the examination of the best specimens of the animals then in the island, thirty points being assumed to constitute perfection. Some years later, this table was revised and settled as follows:

SCALE OF POINTS FOR BULLS.

Article.	Points.
1. Head, fine and tapering	1
2. Forehead, broad	1
3. Cheek, small	1
4. Throat, clean	1
5. Muzzle, fine, and encircled by a light colour	1
6. Nostrils, high and open	1
7. Horns, smooth, crumpled, not too thick at the base, and tapering, tipped with black	1
8. Ears, small and thin	1
9. Ears, of a deep orange colour within	1
10. Eyes, full and lively	1
11. Neck, arched, powerful, but not too coarse and heavy	1
12. Chest, broad and deep	1
13. Barrel, hooped, broad and deep	1
14. Well ribbed home, having but little space between the last rib and the hip	1
15. Back, straight from the withers to the top of the hip	1
16. Back, straight from the top of the hip to the setting on of the tail, and the tail at right angles with the back	1
17. Tail, fine	1

Article.	Points.
18. Tail, hanging down to the hocks	1
19. Hide, mellow and movable, but not too loose	1
20. Hide, covered with fine soft hair	1
21. Hide, of good colour	1
22. Fore legs, short and straight	1
23. Fore-arm, large and powerful, swelling, and full above the knee, and fine below it	1
24. Hind quarters, from the hock to the point of the rump, long and well filled up	1
25. Hind legs, short and straight (below the hocks), and bones rather fine	1
26. Hind legs, squarely placed, and not too near together when viewed from behind	1
27. Hind legs, not to cross in walking	1
28. Hoofs, small	1
29. Growth	1
30. General appearance	1
31. Condition	1
Perfection	31

No prize shall be awarded to bulls having less than 25 points.

Bulls having obtained 23 points shall be allowed to be branded, but cannot take a prize.

SCALE OF POINTS FOR COWS AND HEIFERS.

Article.	Points.
1. Head, small, fine, and tapering	1
2. Cheek, small	1
3. Throat, clean	1
4. Muzzle, fine, and encircled by a light colour	1
5. Nostrils, high and open	1
6. Horns, smooth, crumpled, not too thick at the base, and tapering	1
7. Ears, small and thin	1
8. Ears, of a deep orange colour within	1
9. Eye, full and placid	1
10. Neck, straight, fine, and placed lightly on the shoulders	1
11. Chest, broad and deep	1
12. Barrel hooped, broad, and deep	1
13. Well ribbed home, having but little space between the last rib and the hip	1
14. Back, straight from the withers to the top of the hip	1
15. Back, straight from the top of the hip to the setting on of the tail, and the tail at right angles with the back	1
16. Tail, fine	1
17. Tail, hanging down to the hocks	1
18. Hide, thin and movable, but not too loose	1
19. Hide, covered with fine soft hair	1
20. Hide, of good colour	1
21. Fore legs, short, straight, and fine	1
22. Fore-arm, swelling, and full above the knee	1
23. Hind quarters, from the hock to the point of the rump, long and well filled up	1
24. Hind legs, short and straight (below the hocks), and bones rather fine	1
25. Hind legs, squarely placed, not too close together when viewed from behind	1
26. Hind legs, not to cross in walking	1
27. Hoofs, small	1
28. Udder, full in form—*i.e.*, well in line with the belly	1
29. Udder, well up behind	1
30. Teats, large and squarely placed, behind wide apart	1
31. Milk-veins, very prominent	1
32. Growth	1
33. General appearance	1
34. Condition	1
Perfection	34

No prize shall be awarded to cows having less than 29 points.

No prize shall be awarded to heifers having less than 26 points.

Cows having obtained 27 points, and heifers 24 points, shall be allowed to be branded, but cannot take a prize.

Three points—viz., Nos. 28, 29, and 31—shall be deducted from the number required for perfection in heifers, as their udder and milk-veins cannot be fully developed; a heifer will therefore be considered perfect at 30 points.

In 1866 the Jersey Herd Book was started, and in 1868 the Committee of the Royal Agricultural Society of Jersey called attention in a report to the advantageous results of careful breeding as practised by Mr. Dauncey and others in this country. In a subsequent report in December, 1871, the Committee acknowledged a yearly grant from the State of Jersey of 50*l*., to be applied solely in premiums for bulls, to check the exportation of good animals from the island.

In England, whole-coloured Alderneys, whether dark or light fawn, are decidedly the most esteemed. We believe justly so, and in corroboration of this view we quote from an article by Gisborne in the "Quarterly Review" of 1849 and 1850: "With few exceptions, quadrupeds in a state of nature are self-coloured; and we are not aware of any wild animal whose colours are patchy or glaring. The British wild cattle are of a dingy white with tawny ears. The cattle of mountainous countries, which have been very inaccessible to agriculture, are always of self-colours, black, red, or dun. The queer little cow, which within the memory of man had a pure existence in Normandy and the Channel Islands, and which, being celebrated for the richness of its milk, came to our markets under the name of an Alderney, was fawn colour with tawny ears."

Amongst the herds maintained purely for profit, Mr. Dumbrill's, of Ditchling, near Brighton, is one of the most remarkable. Mr. Dumbrill, who has always adhered to the Jersey breed, keeps one hundred cows, divided into herds of twenty-five each, for the purpose of supplying his wealthy neighbours with butter and cream. In the Brighton market, during the two seasons, there is a demand for the very best of everything in the way of eating without regard to price. In April, 1862, Mr. Dumbrill read before the London Farmers' Club a paper on "Dairy Management," containing practical information of great value to the owners of either trade or fancy dairies.

Another breeder of Alderneys, who bears a name almost classical in the history of agriculture, is Mr. C. H. Bakewell, of Quorndon, near Derby, who has a small but select herd, and which is managed in a profitable manner. His average annual return has been from 220lb. to 240lb. of butter per cow.

This country is well off for breeds of meat-producing beasts, as clearly shown by the articles on Shorthorns, Herefords, Devons, Longhorns, and others. To breed Alderneys with success, in my opinion, no attempt should be made to combine meat-producing with milk-producing qualities. The Alderney breeder, therefore, must be satisfied with an animal almost equal in elegance to a deer, rich in cream, and bountiful in butter of the finest quality. All, however, do not think alike, and an attempt is now being made in a fine herd near London to attain this object.

No doubt one great drawback to the Alderney as a gentleman's cow is that when barren it is often impossible to fatten her, causing thereby considerable loss.

But from this herd last year a cow which had been milked for two years was after three months' feeding, sold in Watford Market by auction for 26*l*. 10*s*. to the butcher; and it remains to be proved whether or not this is an exceptional case.

Heifers kept until three years old before breeding will be larger in frame, but the gain in size is obtained at a sacrifice of dairy qualities, and with increased difficulty in getting them to breed. Alderney heifers should be so managed as to calve at not later than two years and a half old.

Most of the agricultural societies are now offering prizes for Channel Island cattle. The Royal Agricultural Society have recently made classes for both the Jersey and the Guernsey, on the principle that judges who prefer the one may not do justice to the other. This arrangement will, it is to be feared, make the entries in each class very small, particularly so in the Guernsey class, as in this country Guernseys are not numerous. The Bath and West of England Society has of late years secured very good entries for its Alderney classes; and amongst local shows, Essex has been successful in cultivating this truly elegant breed, stimulated perhaps by one or two local breeders, of whom the most successful exhibitor for the past few years, and particularly last year, was Mr. Walter Gilbey, whose bull "Banboy" took first honours at the Royal Agricultural Show, Bath and West of England Show, and the Essex Show at Romford, where also his cows "Duchess" and "Milkmaid" were equally successful.

We may add that the Messrs. Fowler, of Bushey and Watford, Herts, and Southampton, have been the importers into this country of this highly appreciated breed of cattle for nearly a century.

[We think the above arrangement as to points judicious in most respects, but doubt if sufficient points are bestowed upon what may be called the local indications of milk. Thus, a larger udder is most important; but there may be this, and yet no great milk-giving powers. It is necessary not only to be large, but supple, capable of shrinking and wrinkling up after milking, when it should handle soft, flabby, and be much shrunk and wrinkled. The greater the difference between a full and empty bag the better. A fleshy or greasy udder is of uniform texture and firm, resists pressure, and scarcely lessens after milking, and such should be rejected. The same importance is given to the shape, size, and position of the teats as to the form of the udder and prominence of the milk veins. Now, as far as produce goes —putting uniformity and beauty on one side—the shape and size of the teats do not indicate very distinctly; and although, as a general rule, large milk supply is accompanied by large teats standing well out, yet exceptions are by no means uncommon. We cannot agree with "An Amateur Breeder" in his apprehensions that separating the Channel Islands into two classes—which is now done at Bath, the Royal, and Bath and West of England Shows—will be prejudicial to a good entry.

On the contrary, by insuring merit a fair chance of appreciation, we look forward to having these interesting classes better supported than has hitherto been the case.—EDITOR.]

Another authority writes:—

"The only thing required in the engravings after Harrison Weir's excellent drawings is some scale by which to form an estimate of their size, and, though admirable as drawings of cattle, they do not give me a correct notion of the breed. I speak from experience, as I have imported several, and purchased others from the vessel as they landed, and I now possess three undoubted and superior specimens.

"The old breed of Alderney cattle was, I was told, black and white; but when I was a resident (at the time the Prince of Wales was born) there was not one of that colour on the island. The majority were yellow, or yellow and white, sometimes running into fawn, nor do I remember any other colours. The inhabitants insisted on white hoofs and a heart in the forehead, with short horns (alike in size and spread), the large, full, calm eye, and general straightness and symmetry. Ten were shipped for Her Majesty just before I arrived in Alderney by Mr. Gaudion, the judge of the island, and they cost on an average 10*l*. each. The principal points regarded by Mr. Gaudion were those I have named, and the yellow or buttery quality of the inside of the ear; I think he also preferred the white nose and muzzle. Since that time the stud has doubled in value, and I myself paid 20 guineas and 17*l*. for two two-year-old in-calf heifers this spring. I purchased these at the Bear Inn, Weymouth, where they arrive, from twenty to thirty at a time, by the Monday steamers.

"They are usually known as 'Normans' in Hampshire, and hundreds of precisely the same description may be seen browsing in the New Forest, whence they are frequently purchased by unprincipled dealers, after being trained to lead in a halter, washed with soft soap, their horns rasped, scraped, and sand-papered, after which they are distributed about the country. I have more than once seen a labourer thus training a Forest cow for sale.

"The Alderney cow is invariably an excellent butter cow, and, though frequently delicate the first year, and requiring care after first calving, she becomes subsequently inured to our climate. Even heifers require milking in some cases before they calve, to prevent milk fever, but calves dropped in this country are hardy from the first. The cow in full milk will give as much as 12lb. of butter a week, and Inglis says as much as 14lb.; but Mr. Gaudion, who tried the milk by the side of Devons, told me that the latter gave more butter in proportion to the milk. The produce is of a rich golden hue, and it is a common thing with our dairy farmers to keep one or two to "stain the dairy," as high coloured butter commands a sale.

"Of late there has been a great demand for mouse or slate-coloured Alderneys, and they rule at fancy prices. The more dwarfed they are, the better they are liked, and they answer well for paddocks or small pastures, as they never roam or break fences; they may be led or tethered, and they eat no more, or not much more, than a goat. They also do well on garden waste, bran, and cotton cake, or what is called 'hand feeding,' and the skim milk is far better than ordinary milk or the London 'sky blue;' but the butter is too rich for some persons, nor does it keep so well as that obtained from ordinary English cows.

"The Alderney becomes less sightly as it reaches maturity or declines with age. It is most pleasing to the eye as a yearling or two-year-old. The bulls are soon too mischievous to roam at large, but they retain their beauty to the last; whilst the female, especially if a productive milcher, becomes thin, spectral, and misshapen."

CHAPTER XX.

THE BRETON BREED OF CATTLE.

ABOUT TEN YEARS AGO Breton cows were largely introduced into England by Mr. Baker, of Chelsea, with the guarantee that they would suit small families better than the Alderney, from their superior hardiness of constitution and their requiring less food in proportion to the milk given by them. The number imported was quite sufficient to test the qualities of these cattle, and at first sight their almost total neglect in the present day would lead to the impression that their asserted superiority for the purpose of the small dairy was not founded on fact. We believe, however, that these two contradictory positions are reconcilable, and that the good qualities of the Breton cow have been little, if at all, overrated. In the first place, the cattle disease regulations have to some extent interfered with their importation; but we are inclined to believe that the true explanation is that there is always a great difficulty in procuring the male to carry on the breed, for the cow is too small to be safely crossed with any English bull, and consequently either both sexes must be kept by each gentleman who wishes to try the breed, or the milk must soon be lost. Hence it is not surprising that this little cow, so valuable *per se*, should now be rarely met with in England. In March, 1860, the Breton cattle were described in the *Field* as follows, and the description then given holds good to the present day:—"They are various in colour, but all show their distinctive characteristic. They are extremely hardy, and can be well kept on hay and water in any small shed at a cost of 6*d*. a day. One gentleman who has carefully carried out the experiment finds that the keep of one Alderney is equal to four Brittany cows. They appear to have no vice, but are gentle and tame to a degree. They give from four to twelve quarts of milk per day, namely, from four to six quarts with their first calf, and six to eight quarts with their second, &c., and are at their best with the third calf; but in one case a cow only two years old gave with her first calf eight quarts of milk per day. The milk is rich in quality; the result of a test-glass gave 17 per cent. of cream, which is about one-sixth, or say six quarts of milk will give one quart of cream. Their qualities have now been well tested, and their purchasers express satisfaction and admiration. In a few words, they fully answer the description given of them, and it is important to add that in Brittany the lung disease is not known. Lastly, these little cows may be described as a useful and profitable toy. The following is an abstract of a letter from the Messrs. Baker: 'We have received numerous testimonials as to their excellence and good character. They do not vary in the quantity of milk they give, but hold on for some months with a good average quality. They are very docile, easily managed, and invaluable to anyone that has a little grass. They are universally admired for their diminutive size, being from 36in. to 38in. high; and for all small farms from two acres (now so much in fashion amongst ladies) they are specially adapted.'"

Another valued correspondent in 1863 thus alludes to them: "The small holdings and properties in Brittany are just fitted to the diminutive cows and sheep with which they are 'stocked;' and in fields or paddocks about the size of a yacht's mainsail full-sized cattle would be out of all keeping. The name 'brettes' is graphic and amusing; it is like the Scotch 'lassie,' a familiar term of endearment applied to the girls and the cattle alike. The Limousin country resembles Brittany in many respects, but the climate is far hotter, and the land more fertile. Auvergne, further south, bears a much greater resemblance to Brittany, in its soil, and the primitive habits of the people, who are generally believed to be the genuine descendants of the old Italian *coloni*, and to retain unaltered the language, the dress, the agricultural implements and mode of tillage, of the old rustic race described in the Georgics. The Breton cattle are generally tethered, as in the Channel Islands, or else tended by the children while browsing on the roadside. They are very small and lean from long ages of bad keeping, but decidedly pretty and gazelle-looking about the head. They are as much a part of the family as the Irishman's pig, and the cow house is only separated from the keeping room by a bit of baize or canvas. They are doubtless the original stock of the Channel Islands breed, generally known as Alderneys. The latter have had the advantage of richer pasture, pampering with various kinds of food, and a more genial climate; and are proportionately more valuable, although little increased in size. Their yield of milk exceeds in quantity and quality, I believe, that of any other breed, in proportion to their size. The yellowness of the milk and butter, and, indeed, the yellow tinge of the animals' skin, is supposed to arise from bile, produced by excess of soda in the grass, and the practice of close tethering. This is somewhat strengthened by the fact of 'bilious dyspepsia' being among the predominant maladies of the island, though this can scarcely be traced to the same cause. 'The large yield of milk from the island cows,' says the author of 'The Channel Islands' (Allen and Co., 1862), 'and the richness of the milk for butter, are well known. Extreme cases show that from sixteen to seventeen pounds per week of butter have been made from the milk of one cow. They are milked three times a day. Each cow requires about one and three-quarters English acres of grass land, and is fed during winter from the beginning of November on mangold wurzel, turnips, parsnips, and hay.'"

A fuller description of the variety, given by "Marchadour Zaout," a gentleman who has lived in Brittany many years, in reply to the queries of a correspondent, will be found in the following extract from a letter published in March, 1863: "To the first question, as to the height of Brittany cows, I beg to inform J. A. L. that the average height of the pure-bred animal from the sterile 'landes' of the upland districts of this country may be stated at about 36in.—sometimes as low as 32in.—and occasionally, on superior pasture, up to 39in. or 40in. As to their price, that of course depends on quality, state of markets, &c., and expenses of transport, whether in larger or smaller lots, and other circumstances. On an average, however, a useful, well-bred cow, carrying her second calf, could be delivered in England for 14*l*. to 15*l*. To his question, as to the probable result of crossing with a North Devon bull, nothing but actual experiment could afford satisfactory reply. Crossing of breeds is nine times out of ten a mere lottery, and better avoided, unless in such case as that of a given animal, possessing along with certain good qualities sundry defects, which it may be desirable to obliterate or diminish by admixture with some other breed possessing analogous good qualities, and a nearer approach to perfection in those other points in which the other is faulty. Such judicious crossing has obtained for Great Britain the noted excellence of very many of her races of domestic animals; while on the other hand an absurd mania for promiscuous mongrelising has produced in France the lamentable yet ridiculous results that any practical man might *à priori* have assumed as the consequence of mixing together breeds utterly different in nature, or habit, or specialty of production, and without consideration of the new blood infused being adapted or not to the climate, soil, or culture of the district into which the *soi-dit* improved breed is to be introduced. As a few instances among many, I may cite the monstrous puerility displayed, but esteemed as a *chef d'œuvre* by its perpetrators (divers learned *agronomes*), in a cross between the shorthorn bull and a Brittany cow; that with the Swiss bull—an equally wretched one; and of later years, and as yet on a limited scale, the most rational attempt by far that has been made, by means of the Ayrshire. This latter being, like the Brittany, a first-class dairy breed, and analogous in form, habit, and hardiness, may be expected to add to the size of the progeny, without materially, at all events, deteriorating the inherent good qualities of the Brittany breed—hardiness, symmetry, and milchness. My own opinion—and I give it for what it may be worth—is, that this animal, defective only (as a general rule) in size, is susceptible of all the amelioration needful by a judicious selection of parents from its own pure blood, and thereafter good keep of the calf from its earliest age, instead of the all-but-starvation it has now to undergo at the hands of the near-sighted peasant proprietor almost up to the period of its maturity. My advice, therefore, to J. A. L., or any one else, would be to let well enough alone, so far as improvement of this breed goes, by any crossing whatsoever, and, in fact, its smallness of size is a desideratum to many people—for from what other source can the owner of a mere plot of ground obtain a fair supply of home-produced milk and butter? I can argue then nothing in favour of a cross with the Devon, unless it might be that the result of effecting it in an opposite form to that which J. A. L. proposes, viz., the Brittany bull and North Devon cow, would, in the produce, give an increased tendency to produce milk in a smaller and more slender animal. Such effect I have often thought might accrue from such a cross with the Ayrshire cow, and a trial of it might be worth the attention of prize-cattle breeders in Scotland, who may consider the Ayrshire cow as still susceptible of improvement in symmetry and general fineness of structure.

Messrs. Robertson and Co., Eaton Farm, Cobham, Surrey, are now the only importers of cattle from Brittany. To them we are indebted for the following notice by Mr. J. C. W. Douglas, Manoir du Plessis, Chateauneuf du Faou, Finisterre, France, who is well versed in the characteristics of this breed:

"It is questionable if there is a domesticated member of the genus *Bos*, more fully fitted for its situation or surroundings—the right beast in the right place—than that of Brittany, now pretty well known across "ye narrow seas" that separate the 'Grande and Petite Bretagne.' Living on, for the most part, a poor granitic soil among the 'blooming heather,' and the 'lang yellow broom,' which, with the scrubby herbage intermixed, forms their chief, and in many cases only nourishment, they live and they thrive; hardy as West Highlanders, Welshmen, or Kerrys, doubtless they are not; the climate of Brittany, though bleak, and foggy, and ungenial, compared with other provinces of France, is less so than in any of the mountainous districts of the British islands; and then local circumstances modify considerably the calls made on the hardihood of the breed; they are housed at night, and kept indoors in stormy weather, for the wolf, once the terror of our herdsmen, till finally erased from the list of *feræ* by Duncan Cameron, in Badenoch, about 200 years ago, still stalks, a dreaded devastator, over the length and breadth of Brittany. Shelter and safety, however, is about the extent of what the owner's roof affords (and which in many cases is shared with him as fully, and closely, as that of 'Paddy with his pig'). A scanty dole of bog hay, and haply a ration of pounded gorse or furze, is all the cottager and, indeed, the small farmer (who forms the majority of the agricultural class) has to bestow.

"The effect of such treatment shows itself, as might be expected, in the diminutive size of the breed; and in proof of this being an unmistakeable case of cause and effect, is the fact that in every locality to which the breed has been introduced, where the soil is of higher fertility, or the system of culture such as to afford good ordinary forage, in a generation or two, the progeny becomes changed as by an enchanter's wand; the poor slight attenuated frame, with the hinder extremities frequently what is called cat-hamed, and the concomitant indices of early starvation, develops, and expands, and assumes the form of a deep-carcassed, shapely animal, preserving the deer-like head and limbs of its upland progenitor, and surpassing, in the opinion of most people fitted to judge, in its general conformation, as a specimen of what is wanted in dairy stock, the far-famed and justly admired Ayrshires, the breed of milk-cows certainly the hardiest as to short and coarse keep that Scotland produces; and which on this latter point, at least, must yield the palm to the Brittany. As I have constant opportunity of remarking in my immediate neighbourhood—into which the Ayrshire breed was to some extent introduced some years ago, with the worthy intention of affecting improvement in that of the district; but the effort has proved, for the most part, an evident failure; the cross formed does not maintain itself on the keep at the disposal of the poor 'farming bodies,' who form nine-tenths of the bulk of the tillers of the soil, nearly as well as his own unimproved unmingled 'crummie,' whose earliest bite was on the braes, where nothing grows save the coarse 'bent,' and the 'brakken,' and the 'broom,' and which she continues chewing away at till the end, yielding, at an average, her twelve quarts of milk per day, three and a half pounds of butter per week, and in extra cases as much as fifteen quarts of milk in the twenty-four hours, and seven to seven and a half pounds of butter per week.

"I would have wished, in attempting any article on Breton cattle, to have given some information about the origin of the breed. I read a little work on the subject some years ago by a Mr. Bellamy, V.S., of Rennes, and remember being struck by the lucidity of his views about the breeding of cattle generally, and of the pros. and cons. of improvements by crossing, and by selection, and so much the more, of course, approving of them because they agreed with my own; but the book, it appears, is out of print, so I must do without it, and mention merely that in my humble opinion his chapters on the matter of origin led to no positive conclusion, not certainly more so than the theories that Acton and others have advanced on the subject of that of the modern breed of Ayrshires. He (Mr. Bellamy), I recollect, refers to some source for their *fons et origo* as far afield as India, whence in some way or another they were landed at Bordeaux. He inclines also to the belief that to them the Dutch cattle owe their descent; and claims, moreover, as a strong probability, that they were also the progenitors of the Ayrshires. In a country like France, troubled as it has been since 1789, it is not to be wondered at, that rural statistics are sadly wanting, and I fear me that, like many things of their past, the fact of the how and the when of the first appearance of the little black and white cattle of Brittany, 'Launcienne Armorique,' must now remain a mystery, involved in the night of time. Analogy between them and the portly cows of Holland there certainly is not, although there is some similarity of colour; and between them and the Ayrshires there is the similarity of structure and general appearance, that must exist more or less in animals excelling in the production of milk. But I repeat my belief that writers on the subject—and of whom Mr. Bellamy is undoubtedly one of the most enlightened—have generally earned more credit for praiseworthy attempts to solve the question, than the right to be congratulated on having succeeded in doing so.

"Till something, then, new and more reliable turns up, one must be contented, it seems to me, with the fact that Brittany possesses a breed of dairy cattle strongly characterised by the stamp of an old and common descent; and that, thanks to the sturdy stubbornness, apparently inherent among tribes of 'articulate speaking mortals' of Celtic origin, and the small means and blessed ignorance of this particular one in question, their breed of cattle is likely to maintain itself in *statu quo* for a lengthy and indefinite period, so far, at least, as the voluntary system of improvement by crossing goes. Had their improvement been seriously attempted by encouragement of the good selection of breeding animals from the pure stock at the local and regional cattle shows (*concours agricolis*), good might have been done, especially if attention had at the same time been directed to the encouragement of better croppings and root growings. But time and money has been spent by theoretical men in experimenting with different crosses, and thus West Highlanders, Devons, Swiss, Ayrshires, and now latterly Durhams, have been tried. These gentlemen—fancy farmers, and now and again an ambitious peasant proprietor—with more coin and less brains than others of his fraternity, achieved great 'kudos' and cause for speechifying and self-gratulation. While the rank and file of the *petits cultivateurs*, counting as they do ten hundred to one, are left utterly aside. Any practical man of course would be led to smile at, as well as pity, the frivolity of such proceedings, seeing how much might have been done with small outlay under a rational plan of action, and how little with much expense has been really accomplished; and that if indeed the poor little starveling cow, the hardy and shapely milk-giver of the Breton cotter, has, in the past, or is in the future, to escape being swamped by a system of universal mongreling, it is due not to the agricultural intelligence of the period, but to the stolid rough common sense of the native breaker of the stubborn glebe, and who has to live by his trade.

"Having had personally pretty extensive experience in milk-stock of all sorts during some thirty odd years of continued residence in Brittany, I hold the opinion that there does not exist anywhere a breed of dairy cattle naturally so well up to the mark of what a milk cow should be, and so worthy of attention and improvement by the selection of good typical males and females from among themselves, and of course by a system of better feeding. My accompanying sketch gives a good idea of a handsome bull of the breed, I bought him young, and took prizes with him; his father passed into the hands of Messrs. Robertson, of Eaton Farm. The peasant, at the other end of the halter, is a good specimen of a sturdy Breton farmer."

CHAPTER XXI.

THE GUERNSEY BREED OF CATTLE.

By A NATIVE.

FROM TIME IMMEMORIAL the island of Guernsey has been famous for its breed of cattle, and a very just reputation it is, for there are few localities in Europe, and certainly none in Her Majesty's dominions, where a more jealous care has been observed to prevent the mixture of foreign elements. Of course, the isolated position of the island has greatly aided the inhabitants in their endeavours; in fact, we doubt if any but a locality so situated could for so long a period have preserved a breed so intact. The cattle are larger and more valued than even those of Alderney, the name of which is so familiar throughout England. They are exquisitely delicate in form; colours varying from light red to fawn and dun, with a few black, each generally with white intermixed. The head is long and handsome, eye large and prominent, horns gracefully formed. For flesh-giving qualities they are profitable, and for dairy stock they are truly excellent, yielding on the average (if properly fed and cared for) 1lb. of the finest butter per day throughout the year. The size is a fair average, and doubtless the breed would be much larger were it not for the peculiar treatment they have ever been subject to. (The two beautiful animals represented in our drawing were Cloth of Gold, No. 1, and Portia, No. 2, the property of the Rev. J. R. Watson, of La Favorita, Guernsey, who has taken great pains to improve the island breed during his residence in that locality. The bull and cow carried the first prize at the show of the Royal Agricultural Society of England at Bedford.) The farms of the island being limited in size, it is found necessary to tether the cattle, whereby they lose much of that exercise and freedom which would tend to larger growth. They are also by this means too frequently exposed to excessive heat or cold without the possibility of choosing the necessary shelter. Notwithstanding these drawbacks, it is really remarkable how well the animals have always thriven. So great is the demand for this breed that, on an average, seven hundred cows and heifers, with about a dozen bulls, are annually exported.

It is very essential that purchasers of Guernsey cattle should know the character and reputation of those through whom they purchase, inasmuch as the demand has at times induced fraudulent exporters to resort to practices by which other breeds have been only too successfully palmed off on the unwary. The pure Guernseys are chiefly exported through the Messrs. Fowler.

It is interesting to trace from the annals of the island how extremely jealous the native inhabitants have ever been of the reputation of their cattle, and how, when the importation of foreign cattle for breeding purposes has been suggested, the farmers have been alive to the mischief that would ensue, and have not rested until an Act of the Royal Court has been passed to prohibit the introduction of such element and impose heavy penalties on those who may attempt it.

Early in the present century a feeling was prevalent among certain commercial inhabitants of the town of St. Peter Port (not natives of the island) that the importation of cattle from France and other neighbouring countries would tend to improve the local markets by reducing the price of butcher's meat; also that it would enable the islanders to increase their export trade, since the Guernsey breed was in such good repute, and the limited admixture of foreign breeds would but very slightly, if at all, deteriorate the purity, and, while Guernsey blood preponderated, would certainly not detract from the quality. These individuals, therefore, petitioned the Royal Court to repeal the stringent laws which had for many years been in force regarding the importation of foreign cattle, and allow, under certain conditions, the introduction of bulls, cows, and heifers from other countries. Immediately on this movement becoming known, a counter-agitation was set on foot, and we find the inhabitants of nine parishes combining to send up a petition, wherein they set forth that whatever might be the temporary advantage gained to commerce, it would be an ultimate loss to the agricultural interest; it would, moreover, be an unwarrantable innovation regarding one of the distinguishing characteristics of the island, and perpetrate a mischief never to be repaired. Regarding these petitions, the records show that the Royal Court, after having heard the various arguments *pro* and *con.*, and the conclusions of the Crown officers, passed an Act confirming and strengthening the time-honoured custom forbidding the importation of bulls, cows, and heifers from foreign ports, under a penalty of 2*l.* sterling for each animal and its entire confiscation, the said fine to be applied one quarter to the Crown, one quarter to the poor, and the half to the informer. It was further ordered that the said Act should be published in the market, and affixed to the other market laws, that no person may pretend ignorance of the same.

A few years later it would appear that a suspicion or fear was aroused on account of cattle from France having been permitted to be imported for slaughtering purposes, the said cattle being kept alive too long, and allowed to associate with the Guernseys. The inhabitants, therefore, to prevent the possibility of the foreigners being used for breeding, again petitioned the Royal Court, praying that unless stringent measures were adopted to prevent the said mixture of bulls, cows, and heifers of other countries with those of the island, Guernsey cattle would not long be superior to other breeds. In response to this petition we find the Court avowing its appreciation of the value set upon the purity of the Guernsey breed, and their concurrence in the opinion that it was mainly attributable to the unremitting attention and jealous care of the inhabitants that Guernsey cattle were so renowned, and therefore decreed:

1. That all individuals possessing cows, heifers, calves, or bulls imported from France, should, within eight days of the passing of said decree, make a declaration in writing to the high constables of their respective parishes, under a penalty of the confiscation of the said animals, also a fine, at the discretion of justice, not exceeding 20*l.* sterling; and that a copy of said declaration be deposited at the registrar's office.

2. That it be forbidden henceforth to import from France, without special permission of the Court, any bull or bulls, under a penalty of confiscation, and a fine of 20*l.* sterling for each animal imported or otherwise got into possession.

3. That from and after the passing of this decree all cows, heifers, and calves imported into the island direct or indirect shall be killed within four months after their arrival, and foreign calves born in the island, killed within eight weeks after their birth, all under penalty of confiscation of the said animals, and a fine not exceeding 20*l.* sterling.

4. That in the exportation of cows, heifers, calves, and bulls, a declaration shall be made on oath by the proprietor or other person possessing the knowledge that the animals are not only of the island, but that they have not mixed with foreign cattle in any way.

By the published ordinances or Acts of the Royal Court of the island we learn that from time to time the aforesaid decrees have been repeated with stronger and yet stronger force, as for instance in 1823, lest by chance permission of the Court for the importation of foreign cattle might at any time have been unduly or too readily granted, it is ordained that henceforth the said permission be valid only when sanctioned by the president and at least seven of the magistrates; also that foreign cows be not kept alive for more than the allotted four months, even though they be in calf, under the penalty aforesaid, to be imposed on the person who imported them, or in his default the master of the ship that brought them, or in his default the person who had them in his possession.

Frequent attempts have been made at different periods to contravene these laws, but in every instance they have been detected and checked. To the credit of the native inhabitants, however, be it said that such attempts have never been made by them, even though at times some pecuniary advantages might have been gained thereby. The attempted contraventions have always been made by those from other parts who had made Guernsey for the time being their place of residence.

On one occasion it was very plausibly pleaded that certain heifers which had been and were still being imported into Guernsey from France were by no means fit to kill in the four months allotted by the law. Guernsey men, however, were not to be done in this way. They knew that such an excuse meant one of two things—either that the owners wished to keep them for cows, or had an idea of passing them off in England as Guernsey cattle, and they were too proud of their reputation to allow themselves to be thus robbed of it by interlopers. They, therefore, once again appealed to the Royal Court, not only to turn a deaf ear to such excuses, but to forbid for a time the importation altogether, lest a fatal blow should be struck at this their favourite branch of industry. And, in ready compliance, the Court decreed that it is now forbidden to import from France or elsewhere any heifer whatever under the penalty of confiscation and a fine not exceeding 10*l*. for each heifer; also that all masters of vessels carrying cattle are bound, within twenty-four hours of their arrival, to furnish a list to the constables of the parish in which they are landed, under a penalty not exceeding 5*l*. sterling.

We have said at the beginning of our article that Guernseymen will not tolerate admixture into their breed of cattle even from the neighbouring island. In this respect the Guernsey people are much more exclusive than the inhabitants of the larger island of Jersey; and it is this exclusiveness which is their boast and pride. It may be, and indeed is the case, that the breeds of the other islands derive advantage from their admixture with Guernseys—for instance, the old and well-known breed of Alderneys, which is now nearing extinction, have by this means become assimilated to the Guernsey. But, like the Arabs with their horses, Guernsey has ever kept, and boasts of her determination still to keep, her breed of cattle distinct and separate; and hence the law is made equally binding on the importation of cattle from the sister island as from foreign ports.

Acts of the Royal Court have, as we have said, been enforced from time to time, and in no case have the measures been repealed or made less forcible. On the contrary, the system at the present time adopted is even more decided than any we have quoted. For instance, it is now forbidden to import bulls under any pretence whatever; and, moreover, it has become an impossibility for spurious breeds to be exported from the island as Guernsey cattle, since every cow, heifer, and calf imported at Guernsey is immediately branded with the letter F, to signify foreign, and the importer has to deposit the sum of 1*l*., which is forfeited if the animal is not killed within the given time. If, therefore, a purchaser of Guernsey cattle ever supposes himself to have been duped, he has only to ascertain positively whether the animals have ever stood on Guernsey soil. If they have, and are branded with the letter F, they are originally from other places; but if they are free from brand, the purchaser may rely upon it they are the genuine Guernsey breed. We repeat, however, that it is absolutely necessary to ascertain for certain that they have stood on Guernsey soil, for tricks have been practised to avoid the branding, by shipping cattle from France to Guernsey and thence to England without landing them at Guernsey at all, or at least only landing to pass from one vessel to another.

The most satisfactory method for gentlemen desirous of being possessed of Guernsey cattle is to visit the island for themselves. A splendid choice would be afforded them at the time of the show, which is held annually on Whit-Tuesday, when from 170 to 200 head of pure Guernseys may usually be seen; and no one could witness the sight without being impressed with the beauty and quality of the breed.

There is another show held on the 20th of December in each year for fat cattle.

A good idea of the class of animals to be seen at this exhibition may be gathered from the following list of weights of beef at the 1872 show, it being borne in mind that the quotations given are Guernsey weight, 102lb. of which are equal to 112lb. English. It is, moreover, worthy of note that from the age of two years till within six months of their death the oxen are doing the work of the horse at the cart and the plough.

By whom fed.	Age.	Weight.
Thomas Hocart	8¾ years	1199lb.
Samuel Best	4½ years	966lb.
P. Hocart	7 years	1152lb.
J. Le Page	6¼ years	1257lb.
John Corbin	5 years	916lb.
J. Le Page	5¾ years	830lb.
Col. Feilden	6 years	908lb.

In conclusion, as our article may possibly lead to the Guernsey cattle being even more than ever in request, we will just observe that purchasers in England must not be too hasty in condemning us or the Guernseys if they do not find the cattle at once turning out all we that have described. In this matter, like all others, a fair trial is necessary. With cows and heifers especially, the change of climate and pasture may have a checking effect at first; but as they become naturalised to their new home they will soon prove themselves true to the character they have so long and so justly borne.

In the management of Guernsey cows the islanders make it a special study to supply that kind of food, and otherwise subject the animals to that method of treatment, which tends to promote good milking qualities. Flesh-making is quite a secondary consideration with them, so long as good quantity and quality of milk and butter are forthcoming.

They have their special treatment for special seasons. For instance, in spring and summer the cows are fed on clover, lucerne, and grass, care being taken that neither is of too mature a growth, and that only a given quantity of each is supplied. Clover and grass, when not too ripe, produce abundance of milk; lucerne also is beneficial, but not in too great quantity, otherwise it is apt to kill the delicate flavour of the butter. If these foods are supplied in too ripe condition they tend to make flesh instead of increasing the milk.

Those who take pains with their cattle are careful that the cows feed when the sun is shining, and place them in the shade as soon as a sufficient quantity of food has been taken, as allowing them to remain in the heat after feeding tends to reduce milk.

In winter the cows are fed on hay, straw, carrots, turnips, and mangold wurzel. The greatest quantity of milk is produced from food in the following ratio: first, carrots; second, common turnips; third, mangold wurzel. Other roots, such as parsnips and Swede turnips, are avoided where practicable, inasmuch as they produce flesh rather than milk. Care is also observed by those who take pride in their cattle to house them at night and in rough weather, allowing them open air and exercise only when the sky is clear.

Disease among cattle is a thing comparatively unknown in Guernsey. The majority of premature deaths occur from calving or milk fever, but these are very rare. Every precaution is taken to prevent diseased cattle being imported, and equal precaution is observed with regard to the native cattle coming in contact even with the healthy importations.

Guernsey heifers are frequently in calf at the age of two years, and, having calved, they yield milk to the age of fifteen, and often beyond that time, being dry at intervals to the extent of only from three to six weeks each time. It is sometimes urged that heifers are permitted to breed too young; but the experience of many, if not most Guernsey breeders, is that the cows are not a wit the worse for it.

In England there exists a strong prejudice against cow beef, but the method of feeding in Guernsey is such as to render the flesh of the cow as delicate as the heifer, that is, so far as the prime joints are concerned. Flesh-making is avoided until the animal has run dry, and its time has come for slaughter; then attention is turned to the clothing of bone and sinew with new flesh, which, when killed and brought to table, is found to be as tender and delicious as can be desired.

INDEX.

A.

Abortion, sympathy in .... *page*	10
Agricultural societies of Kerry as improvers of Kerry cattle	137
ALDERNEY CATTLE, by an Amateur Breeder	139
,, cream and butter producers	139
,, value for town dairies, or gentlemen's parks	139
,, no resemblance to the Normandy cattle	139
,, Mr. Fisher Hobbs as to the probable origin of	139
,, bred in England superior to native	140
,, prices of at Mr. P. Dauncey's sale in 1867	140
,, improved by rules of Jersey Agricultural Society	140
,, scale of points for bulls	140
,, ditto, cows and heifers	141
,, herd book of, in 1866	140
,, value of whole colour	142
,, Mr. Gisborne on colour	142
,, Mr. C. H. Bakewell's at Quorndon ...	142
,, difficulty of feeding	143
,, age to breed at	143
,, at English shows	143
,, importance of local indications of milk in	143
,, character of cattle in the islands, 1840	144
,, Mr. Gaudion's herd of	144
,, early delicacy	144
,, large yield of butter	144
ANGLESEA CATTLE, by M. Evans	129
,, estimates as to exports	130
,, allied to Pembrokes	131
,, pure black colour	131
,, coarser fore quarters and better hind quarters than Pembrokes	131
,, milking properties neglected	131
,, very valuable for grazing	131
,, character of	131
,, in great demand by drovers	132
,, successfully bred in Carmarthenshire, and Merionethshire	132
,, Bakewell's opinion of	132

Anglesea, not suitable for crossing .... *page*	132
,, instance of albinos	123
,, early history of	129
ANGUS, OR ABERDEENSHIRE CATTLE, by Scotus ...	101
,, distinct from Galloways	102
,, cross with Galloways not successful	102
,, districts in which bred	102
,, present breeders of	102
,, points and character of	102
,, importance of purity	103
,, cross made with Shorthorn bull	103
,, black colour of general, but not universal	104
Aptitude to feed according to breed	2
Artificial food on grass	18
Ashes for yard bottoms	41
Asphalte floors for cattle sheds	40
Ayr, county of, districts	105
AYRSHIRE CATTLE, by G. Murray	105
,, improvement of, from 1780	106
,, Aiton on	106
,, crossed with Teeswater	106
,, milking properties of	106
,, statistics of milk yield	107
,, points of	107
,, characteristics of milking properties	108
,, American herd book of	108
,, cows, price of	109
,, cattle, system of letting cows	109
,, mode of feeding	109

B.

Bailey Denton on covered yards	38
Barley, analysis and value of	33
Beans, ditto	33
Bowley on breeding	10
Boxes and stalls, comparative advantage of	44
Breed, age at which to	9
Breeding and feeding	21
Breeding herd, management of	7
Breeding or feeding, a question of soil	18
BRETON CATTLE, by J. C. W. Douglas	149
,, cows, introduced by Mr. Baker	146
,, described in *The Field*, 1860	146

INDEX.

Breton Cattle suitable to cultivation of Brittany
  page 147, 149
  „ commonly tethered 147
  „ supposed parents of Alderneys 147
  „ yellow skin supposed to arise from
    bile 147
  „ described by Marchadour Zaout 148
  „ small size distinctive and valuable
    character 148
  „ food of 149
  „ diminutive size attributed to hard keep 149
  „ origin involved in mystery 150
  „ failure of attempts to cross 151
Building materials 41
Buildings, variety of plans....................... 42
  „ details of 43
Bulls, value of good blood 2

C.

Calf, management of 11
Calves, summer feeding of 12
  „ winter management of 14
  „ condimental foods unsuitable for 15
  „ Mr. H. Buck's management of 16
Calving, preparations for 10
  „ cows, management of during 10
Carbolic Acid, value for cleansing skin 13
Cattle, early history of 1
Cheese, produce increased by palm-nut meal 53
  „ influence of food 58
  „ factories in Derbyshire 58
  „ making, Cheddar process 58
Climate determines value of cattle 5
Colour, value of 3
  „ deterioration of 3
Condiment, recipe for 34
Cotton cake, inferior, injury from 29
  „ analysis of 28, 29
Cotton seed cake decorticated.................... 28
  „ undecorticated 29
Covered yards, when desirable 35
  „ economize litter...................... 36
  „ manure more uniform 36
  „ not adapted for breeding animals...... 37
  „ cost of, Bailey Denton 38

D.

Dairy cow, average yield from 52
  „ advantage of artificial food for 53
  „ wanting in quality 2
  „ method of feeding 49
  „ kept on arable land 49
  „ management of before calving 50
  „ Mr. Horsfall's experience 51

Dairy, temperature and regulation .... *page* 56
  „ construction of 56
  „ furniture of 57
  „ Mr. Horsfall's 57
  „ good water supply 57
  „ management, Schwartz 59
  „ deep can and low temperature system
    at Hofgarten 62
Darwin on colour of early cattle 123
Davey, Captain J. T., on Devons 78
Derby factory, proportion of milk to cheese 107
Devons, by Captain J. T. Davy 78
  „ common origin with Hereford and
    Sussex 78
  „ compared with Sussex 79
  „ influence of climate and food on form
    and character of..................... 79
  „ description of 80
  „ economical meat makers 80
  „ value of in America (Mr. Steinmetz) 80
  „ value for draft purposes................ 81
Douglas, J. C. W., on Breton cattle 149
Duckham, T., on Herefords..................... 72

E.

Exercise, importance of 4

F.

Farm buildings, deficiency of in South Wales ... 124
Fatting cattle, sketch of 19
  „ cost of 20
Fatty degeneration and the Smithfield Show of
  1873 8
Feeding, irrational mode of...................... 22
  „ cost of and return from................ 34
Food, cooked *versus* raw 55
Fowler, J. K., on breeding 8
Fulcher, T., on Norfolk and Suffolk red polled
  cattle 92

G.

Galloway Cattle, by G. Murray 97
  „ native home of 97
  „ characteristic features of 97
  „ black colour of, not invariable 97
  „ great importance of in early times ... 98
  „ important southern trade with possible
    progenitors of English counties polled
    breeds 98
  „ not injured by Irish cross 100
  „ bad character as milkers due to mis-
    management 100
  „ produce small but rich 100
  „ improved modern management of 101

INDEX.

Galloway Cattle, crossed with Ayrshires ...... page 101
GLAMORGAN CATTLE, by M. Evans 117
  „ now rapidly being superseded 114
  „ improved by Normandy cattle, 12th century 118
  „ Devon influence on 118
  „ great value in 18th century 118
  „ bred at Windsor by George III. 118
  „ decline of, ascribed to breaking up pasture land 119
  „ modern breeders of...................... 119
  „ description of the Treguff breed 119
  „ great value for dairy purposes 119
  „ good workers and slow feeders 120
  „ description of, by Youatt and Martin 120
  „ former habit at Monmouth and Gloucester............................... 120
  „ only pure herd at Badminton 120
  „ description of Badminton herd, by Mr. J. Thompson 121
GUERNSEY CATTLE, by a Native 152
  „ high reputation of......... 152
  „ jealousy as to purity..................... 152
  „ character and features 152
  „ illustration of Bedford prize cattle ... 152
  „ yield of butter 152
  „ numbers annually exported 152
  „ large shows on Whit Tuesday and December 20th 155
  „ weight of fat cattle at show in 1872 156
  „ require acclimatising 156
  „ management of in the Island 156
  „ winter and summer feeding of 156
  „ freedom of disease in the island 157
  „ breeding age 157
  „ excellence of cow beef 157
Guernsey Royal Court prohibitions as to importation of cattle........................... 153
  „ decrees 154

H.

Heasman, A., on Sussex cattle 88
HEREFORDS, by T. Duckham 72
  „ fine quality of flesh 72
  „ success of Smithfield shows 73
  „ different varieties of 73
  „ general character of 73
  „ herd book of 73
  „ localities of 5
  „ spread of, at home and abroad 74, 75
  „ success of, in Australia 76
  „ moderate milking properties of 73
Hofgarten dairy arrangements 61
Horsedung as food for cows 60
Horsfall on dairy cows............................ page 51
House feeding for milk cows 64

I.

Indian corn, analysis of 51
  „ pig feeding on, Lawes 32
Irish cattle, improvement of 2
  „ method of removing horns 99
Irish polled cattle, formerly imported into Scotland................................ 99
Italian rye grass for dairy cows 50

K.

KERRY CATTLE, by R. O. Pringle 134
  „ middle horn variety 135
  „ points of 135
  „ Dexter variety 135
  „ points of Dexter cattle 135
  „ true Kerry, as distinguished from Dexter 135
  „ crossed with West Highland............. 136
  „ neglect of breeders 136
  „ bear confinement 136
  „ dimensions of Alderman Purdon's ... 136
  „ the poor man's sort 136
  „ yield of butter and milk 136
  „ quality of flesh 137
  „ modern breeders 137
  „ prize bull at Dublin show 137
  „ English breeders of 137
  „ prices of................................... 137
  „ Mr. Bogue, salesman of................. 138

L.

Lentils, analysis and value of........................ 33
Linseed oil cake, analysis, variety and value of 27
  „ objections to use of 27
Locust bean, valuable as a mixture 34
LONGHORNS, past and present importance of 83
  „ original home of 83
  „ early improvers of 83
  „ Bakewell's herd of...................... 84
  „ description of the bull Shakspeare ... 84
  „ prices at Mr. Fowler's sale 85
  „ description of, by Mr. Marshall 85
  „ ousted by Shorthorns 85
  „ valuable milkers 86
  „ in Derbyshire............................ 86
  „ present herds of......................... 86
  „ present characteristics of 86

M.

Malt, valuable as a mixture 33
Mangers valuable, racks useless 40

INDEX.

Manure, manufacture and preservation .... *page*	45
" in covered yards superior	46
Manure made in open yards	46
" preservation of, in heaps	47
McDougall's disinfecting powder	46
Milk, average yield of	52
Milk farm, management of	63
Milking properties injured by overfeeding	7
Mona, explanation of, by Rev. R. Ellis	130
Morgan Evans on Glamorgan cattle	117
" on Pembrokeshire cattle	8
" proposition as to Welsh black cattle herd book and stock breeding company	127
Murray, G., on Ayrshire cattle	105
" on Galloway cattle	97

N.

New Forest cattle similar to Alderneys	144
NORFOLK AND SUFFOLK POLLS, by T. Fulcher ...	92
" Mr. Marshall's description of	92
" difference in type of, and Scotch polled cattle	93
" Mr. George's herd of Eaton	93
" principal herds in Norfolk	94
" " Suffolk	94
" characteristics and qualities of	94
" recent improvements of ....,....	84
" Mr. J. K. Fowler's opinion of	94
" red polled cattle, resemblance to Prince Leichtenstein's	95
" herd book of, by H. F. Euren	95

P.

Palm nut meal (Smith's), analysis and value of...	31
Peas, analysis and value of	33
PEMBROKESHIRE CATTLE, by Morgan Evans	122
" antiquity of the breed	122
" early colour uncertain	122
" opinion on, by Youatt	123
" suitable for exposed situations	123
" more suitable to their district than Herefords or Shorthorns	125
" can live out all winter	125
" character of	125
" character of horn desirable	126
" long curling hair, not indicative of feeding properties	126
" meat and milk excellent	126
" tradition of improvement by Devon bulls	126
" similar story as to Herefords............	126
" experience of the writer opposed to crossing	127

Pembrokeshire Cattle, herd book of, vol. i., 1874 *page*	127
" errors in popular descriptions	128
" localities in which they are bred	128
Pringle, R. O., on Kerry cattle	134
Proportion of stomach and intestines indicates nature of food required	26
Pulpers, Messrs. Hornsby and Sons'	23
Pulping machine, value of	22
Purity, influence of....................................	3

Q.

Qualify, importance of, in breeding	7

R.

Rape cape, analysis and value of	30
Ruck, Henry, on management of calves............	16

S.

Scotus on Aberdeen cattle	101
Shedding in grass field, improvement of............	39
SHORTHORNS, by J. Thornton	67
" universality of	67
" rise of	67
" hint as to forming herd...................	4
" influence and value of	4
" C. Colling's sale, 1810	68
" characteristic features of	68
" early breeders of	68
" systems of breeding	69
" herd book of	69
" comparative past and present value of	69
" prices of, at New York Mills............	70
" preponderance of, at Royal shows......	70
" value for milking purposes	70
" value of, for crossing, *vide* Irish cattle	70
" progress of, in America and Colonies	71
Skin, importance of cleansing	13
Soiling system, economy of............................	19
South Wales, soil and climate suitable for breeding	124
Straw and hay, comparative value of	25
Straw, feeding value of	24
Substitutes for straw	39
SUSSEX CATTLE, by A. Heasman	88
" original value for draft purposes	88
" contrast of modern and ancient type	89
" dimensions of the Burton ox............	89
" general features of........................	89
" improvement of, by Mr. E. Cane	89
" description of prize steer at Smithfield show, 1867	90
" herd book of	91

T.

Thompson, Mr. J., on old Gloucesters ...... *page*	121
Thornton, John, on Shorthorns	67

W.

WEST HIGHLAND CATTLE, by John Robertson ...	110
,, original breed of the West	110
,, originally from the mountainous districts	111
,, characteristics of	111
,, loss of, in Rannoch from starvation ...	111
,, superseded by sheep stock...............	111
,, quality of beef, slowness of feeding ...	112
,, at Falkirk Tryst........................	112
,, anecdote of "The Baron"................	112
,, English trade declined	113
West Highland Cattle in the western islands *page*	113
,, herd of Messrs. Stewart................	113
,, Mr. McDonald's herd in Uist	113
,, Mr. Malcolm's at Poltalloch	113
,, superior in Argyle.....................	114
,, Mr. Fraser's Faillie "stots"............	114
,, Lord Breadalbane's herd of	114
,, herd of Blair Athol	114
,, high prices of, at Bredalbane sale......	115
,, management of	115
,, scanty but rich milkers.................	116
,, crossed with Shorthorns	116
White animals, delicacy of	3
,, liability to parasites	3
Wilde, Sir W. R., on ancient cattle in Ireland ...	134
,, on four extinct races	135
Winter feeding, expensive necessity	19

GUERNSEY CATTLE, THE PROPERTY OF THE REV. J. R. WATSON.

A BRETON BULL, THE PROPERTY OF J. C. W. DOUGLAS, ESQ.

BRETON CATTLE.

ANGLESEA CATTLE.

PEMBROKESHIRE OR CASTLEMARTIN CATTLE.

GLAMORGAN CATTLE

WEST HIGHLAND CATTLE.

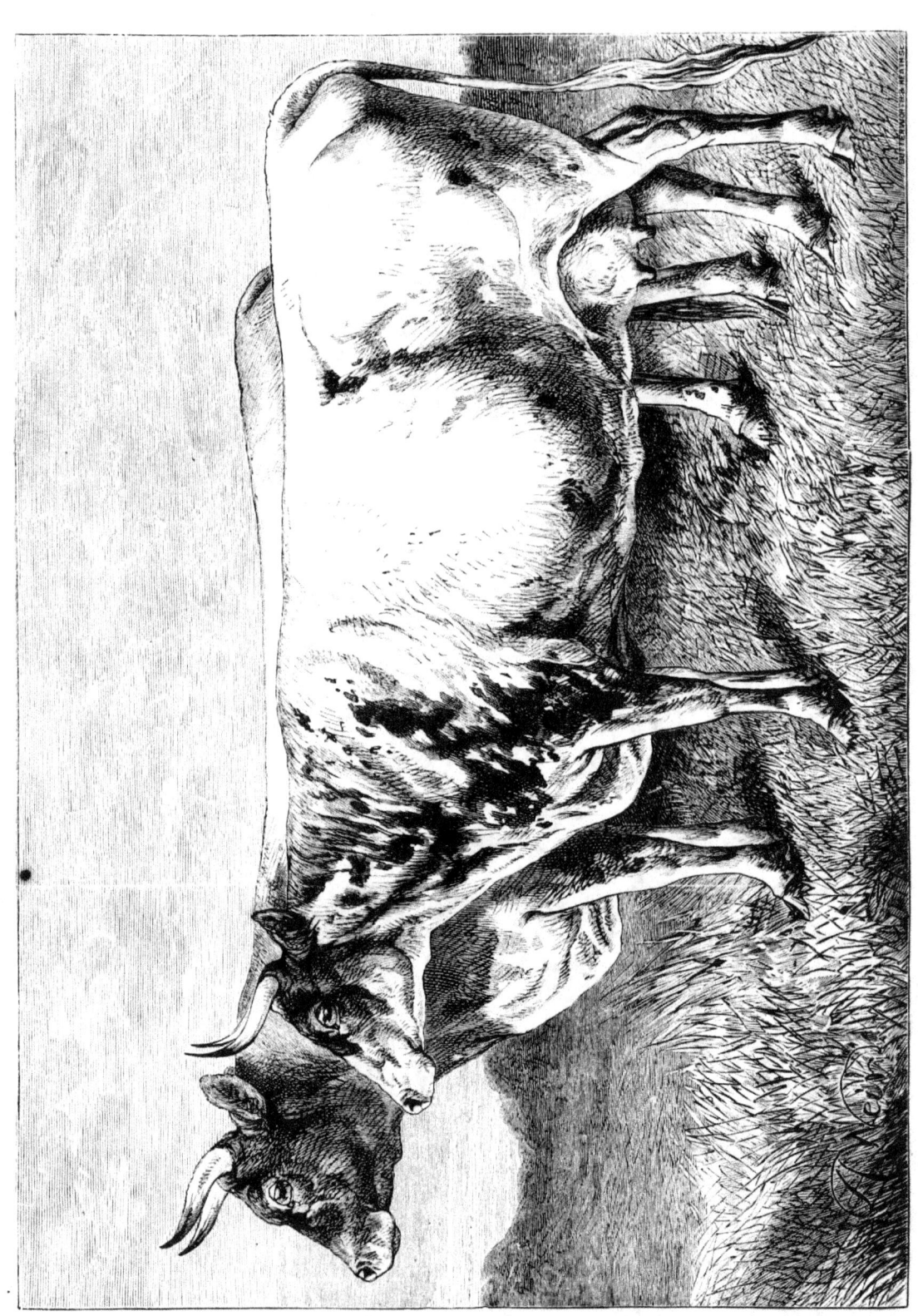

AYRSHIRE CATTLE.

POLLED ANGUS OR ABERDEENSHIRE CATTLE.

SCOTCH POLLED CATTLE

NORFOLK AND SUFFOLK RED POLLED CATTLE.

SUSSEX CATTLE.

LONGHORN CATTLE.

DEVON CATTLE.

HEREFORD CATTLE.

SHORTHORN CATTLE.

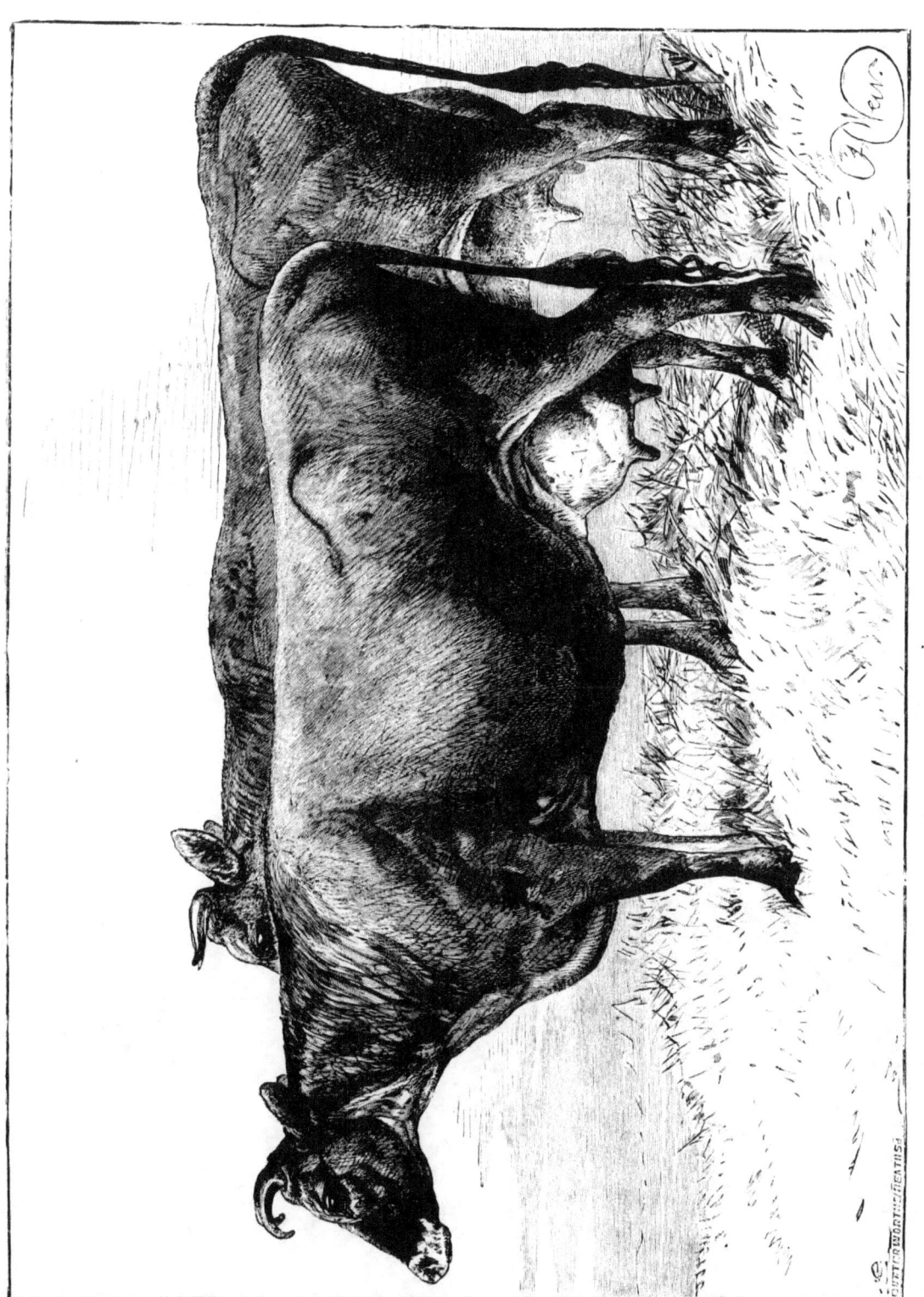

ALDERNEY COWS, THE PROPERTY OF WALTER GILBEY, ESQ.

www.ingramcontent.com/pod-product-compliance
Lightning Source LLC
Chambersburg PA
CBHW062215220526
45471CB00009B/3206